GOVERNORS STATE UNIVERSITY LIBRARY

3 1611 00137 3775

DATE DUE

Applications of Artificial Intelligence in Chemistry

Hugh M. Cartwright

Physical Chemistry Laboratory, University of Oxford

GOVERNORS STATE UNIVERSITY
UNIVERSITY PARK
IL 60466

OXFORD   NEW YORK   TOKYO
OXFORD UNIVERSITY PRESS
1993

Oxford University Press, Walton Street, Oxford OX2 6DP

Oxford New York Toronto
Delhi Bombay Calcutta Madras Karachi
Kuala Lumpur Singapore Hong Kong Tokyo
Nairobi Dar es Salaam Cape Town
Melbourne Auckland Madrid
and associated companies in
Berlin Ibadan

Oxford is a trade mark of Oxford University Press

Published in the United States
by Oxford University Press Inc., New York

© Hugh M. Cartwright, 1993

All rights reserved. No part of this publication may be
reproduced, stored in a retrieval system, or transmitted, in any
form or by any means, without the prior permission in writing of Oxford
University Press. Within the UK, exceptions are allowed in respect of any
fair dealing for the purpose of research or private study, or criticism or
review, as permitted under the Copyright, Designs and Patents Act, 1988, or
in the case of reprographic reproduction in accordance with the terms of
licences issued by the Copyright Licensing Agency. Enquiries concerning
reproduction outside those terms and in other countries should be sent to
the Rights Department, Oxford University Press, at the address above.

This book is sold subject to the condition that it shall not,
by way of trade or otherwise, be lent, re-sold, hired out, or otherwise
circulated without the publisher's prior consent in any form of binding
or cover other than that in which it is published and without a similar
condition including this condition being imposed
on the subsequent purchaser.

A catalogue record for this book is available from the British Library

Library of Congress Cataloging in Publication Data
Cartwright, Hugh M.
Applications of artificial intelligence in chemistry / Hugh M. Cartwright.
p. cm. — (Oxford Chemistry primers; 11)
1. Chemistry—Data processing. 2. Artificial intelligence.
I. Title. II. Series.
QD39.3.E46C37   1993   540'.285'63—dc20   93–15892
ISBN 0 19 855737 X (Hbk)
ISBN 0 19 855736 1 (Pbk)

Typeset by the author
Printed in Great Britain on acid-free paper by
the Bath Press

QD 39.3 .E46 C37 1993

Cartwright, Hugh M.

Applications of artificial
  intelligence in chemistry

312520

Series Editor's Foreword

Oxford Chemistry Primers are designed to provide clear and concise introductions to a wide range of the topics that may be encountered by chemistry students as they progress from the freshman stage through to graduation. The Physical Chemistry series is designed to contain books easily recognized as relating to established fundamental core material that all chemists will need to know, as well as books reflecting new directions and research trends in the subject, thereby anticipating (and perhaps encouraging) the evolution of modern undergraduate courses.

In this, the first Physical Chemistry Primer, Hugh Cartwright has produced a stimulating and easy-to-read account of the use of Artificial Intelligence in Chemistry. This is an emergent topic which is expected to facilitate the study of new types of problems within chemistry. This Primer will interest all students (and their mentors) who wish to see how modern computational approaches can transform the methodology of science.

Richard G. Compton
Physical Chemistry Laboratory, University of Oxford

Preface

The tools of science are changing. Within the past decade, computers have become as crucial in science as chemicals or spectrometers. Recently there have been even more far-reaching developments: artificial intelligence has spread to the laboratory, where it promises to have a profound effect upon scientific practice.

Several powerful techniques, which emulate human thought and reasoning, constitute the area of artificial intelligence, one of the most fascinating and exciting in science. Problems in this area abound. They may appear complex and intractable when tackled using traditional methods of solution, but artificial intelligence methods are solving them with growing success.

As these new intelligent methods expand in power, scientists are turning to them increasingly to help both in everyday tasks, and, more dramatically, to develop new scientific theory and understanding. This book provides a non-mathematical introduction to this intriguing subject and its potential. In it we investigate the nature of Artificial Intelligence, how it is applied, and how in the coming decades it will assume an increasingly fundamental role within science.

Many people contributed to this primer. My students Stephen Harris, Nats Imai, and Rob Long, and my daughter Jenny, read drafts and made many useful suggestions. John Freeman's excellent diagrams are a vital part of the book. The series editor Richard Compton and OUP provided expert editorial help. Above all, my wife Susie and daughter Jenny offered constant support and encouragement. I am grateful to them all.

Oxford
June 1993

H.M.C.

Contents

1 Artificial intelligence

1.1 Introduction

A machine that could complete *The Times* crossword or remove a human appendix unaided would undoubtedly be intelligent. At present such machines exist only in science fiction, but already laboratories around the world are using 'Artificial Intelligence'. What does this mean? Why is artificial intelligence important to science? And how can it be used in the laboratory? These are the central questions that this book sets out to answer.

Artificial intelligence (AI) has evolved over forty years. For most of that time development was slow-moving and academic. Work centred on game-playing and puzzle-solving machines, not because applications in these areas were of practical importance, but because tasks of more widespread interest were far too complex for the computers then available.

Table 1.1 Some early AI programs

Program	Purpose
LOGIC THEORIST (1956)	To solve problems in propositional logic
BASEBALL (1961)	To answer questions about baseball
STUDENT (1964)	To solve algebraic problems
DENDRAL (1965 onwards)	To analyse mass spectra of organics
SHRDLU (1972)	To manipulate coloured blocks
MYCIN (1972 onwards)	To select anti-microbial agents

Within the last decade, however, as computer power has increased, the field has expanded in a most dramatic fashion. The enormous potential of AI is now starting to be recognized and harnessed, not just by computer scientists, but also by those within the physical and life sciences, and beyond. As we shall see, this has significant implications for chemistry.

We shall explore in this primer how AI is applied to those areas of chemistry which offer great scope for investigation by intelligent methods. The early development of a new field in science is often a period of rapid and

Recent changes in computer power and size were quite unforeseen just a short time ago. In March 1949, a popular science magazine predicted that in the distant future computers might have 'only 1000 vacuum tubes and weigh only one and a half tons'.

exciting growth, and the evolution of AI is now at that stage; as a result this is a particularly fascinating area of work.

Artificial intelligence is, of course, quite different from chemistry, yet in a very real sense it is an experimental subject: we are still learning how to use its methods and interpret its predictions. As a result, the application of AI to science is full of investigation, as we try to understand how its techniques can be used.

AI is a young and largely unexplored subject. Despite its short history, however, its potential is unmistakable: AI will be a central part of the chemist's arsenal within two decades.

1.2 What is artificial intelligence?

AI is an attempt to reproduce intelligent reasoning using machines.

Some very unmachine-like entities can act intelligently. In one of the earliest studies of AI, a set of 288 matchboxes filled with coloured beads (and known affectionately as MENACE) learnt to play an expert game of noughts and crosses (tic-tac-toe). However, a matchbox collection is not a prerequisite for an interest in AI. All practical work nowadays is carried out without matchboxes, and in this book we consider only methods which use more sophisticated tools.

The word 'machines' is used here in a sense that is much broader than its everyday meaning; almost any inanimate mechanism which can complete a task qualifies as a machine.

In scientific work our prime tool is the computer, which reasons rapidly and reliably; it is an essential part of all practical applications of AI. Nevertheless, although there are strong links between AI and computer science, a knowledge of computers is not necessary to understand the principles of AI, or how those principles apply to science. Nor do you need to be a computer expert to understand the contents of this primer; we shall often refer to computer programs, but always in a non-technical fashion.

Indeed, the relatively non-technical nature of AI helps explain why it has such broad appeal. Everyone knows how to reason after a fashion, and most of us have some common sense. Since it is the purpose of AI to emulate human thought and reasoning, and common sense is just an expression of this, each of us already knows something about the subject. In a way, taking advantage of AI is not unlike trying to teach a machine to use some common sense.

However, to suggest that to make a computer 'intelligent' we merely need to instil in it a bit of common sense is to simplify things a little. Common sense is intangible; it certainly is not something that we can easily explain to computers, or even to people. Conventional ways of telling computers what to do are poorly suited to the goal of helping them to learn, and if a computer is to behave intelligently, not simply as an obedient but dumb calculator, we will have to re-think how computers are given their instructions, and what those instructions are.

It is the purpose of Chapters 2–4 to explain how this is done. Computers *can* be persuaded to act intelligently, and thereby solve problems which may resist solution using dumb methods. In these chapters we shall discover the striking differences between the way scientific problems are tackled using conventional methods, and the way they are solved using the alternative methods of AI.

Before we look at individual AI methods, it is useful to consider what kind of problems AI might help us to solve. Not every scientific problem is a candidate for this type of approach, and you might wonder whether scientific problems that are soluble using AI have any identifiable characteristics by which we might recognize them. As we consider this question, we shall gain a first insight into the reasons why AI techniques are so powerful.

1.3 The scope of artificial intelligence

Domains

A remarkable feature of the human mind is the range of topics about which it can successfully reason. According to the definition given at the start of the previous section, AI tries to achieve something close to human reasoning; we can work in an almost unlimited variety of different areas or **domains**, so it seems that *every* problem of interest to scientists must be a problem in AI.

A domain is an area of expertise in which an AI program, or a person, operates.

Unfortunately, no AI techniques yet have the power to solve scientific problems of arbitrary complexity; they operate only in restricted domains: that is, a particular method is required to solve a particular type of problem. AI is at the moment quite unable to transform a computer into a machine that behaves like a person, and it cannot yet show us how to develop all-powerful machines which can solve any problem. In other words, we cannot just point an intelligent machine at any scientific problem we want solved and say 'get on with it'.

On the face of it, this is disappointing news. However, although AI methods are presently restricted to particular domains, it would be wrong to conclude that there are few problems in science suited to AI, or that AI methods are of only peripheral importance. In fact, quite the opposite is true.

Particular types of scientific problem do require an appropriate AI method for efficient solution, but these methods are available already, and are growing in sophistication and power. Not only are there numerous problems well suited to AI, but many of these are essentially unsolvable by any conventional method. AI problem-solving is limited at present mainly by the need to understand better how AI methods work, and to a lesser extent by limitations in computer power. As our understanding grows, the use of AI will expand to encompass all areas of science.

AI problems

In this book we discuss those AI methods which appear to offer the greatest potential in solving scientific problems: artificial neural networks, expert systems, and genetic algorithms. The range of problems amenable to these techniques is considerable, but not all-encompassing, so we must distinguish between problems suited to AI methods of solution and those more appropriately solved by conventional techniques. This division into AI and non-AI problems is surprisingly straightforward.

Scientific AI problems are characterized by one notable feature which we will use as their definition:

A problem in artificial intelligence is one which is so complex that it cannot be solved using any normal algorithm.

A 'normal algorithm' in this definition is a precisely-defined sequence of instructions to a computer which will deliver the answer to a scientific problem in a reasonable time (perhaps an hour on a fast computer).

Far from being a helpful definition, this sounds rather alarming. It seems to suggest that all the easy problems in science can be solved using conventional methods, leaving behind for AI the residue of those which are well nigh incomprehensible. Fortunately this is not the case.

The problems which conventional methods fail to solve fall into two categories: they may be computationally intractable or conceptually obscure. The obscure problems are those which are so convoluted or poorly understood that it is unclear how to even start to analyse them. Neither AI nor conventional methods make much impact at present on such problems, and we shall not consider them further.

Computationally intractable problems are quite different. These are exactly what AI methods thrive on – indeed that is the implication of the definition of an AI problem given above. A computationally intractable problem may be hard to solve, but that does not mean it must be difficult to understand.

A computer working since the universe began to investigate the sequences of moves in chess would by now have been able to investigate only a fraction of all possible games.

Consider the task of teaching a computer to play chess. This is a problem in AI: purely numerical or exhaustive search algorithms are incapable of play at the highest level. And yet, though playing chess at grandmaster level is a challenge beyond all but a few of us, the rules of chess are straightforward. The problem is computationally demanding, yet still easy to understand.

Similar situations arise in science. Prediction of the stereochemistry of proteins is a thorny computational problem, but we can understand in simple thermodynamic terms why a protein wants to adopt a geometry which minimizes its free energy. The selection of optimum parameters for HPLC analyses, the computer elucidation of new synthetic routes, or the control of a large-scale industrial plant are computationally demanding, yet in every case the problem is far from being incomprehensible.

So, although it may seem odd that we define scientific AI problems in terms of how difficult they are to solve, rather than in terms of their nature, we see that complexity of solution is at their very heart.

In view of this definition, it is possible to immediately separate out many problems that are best tackled by non-AI methods. Simple problems can be solved using simple methods. It would be inefficient to use AI to calculate a least squares fit of a straight line through a set of data points, for example: purely numerical methods are far faster. Equally, though calculation of the energy levels of a molecule of ascorbic acid is not trivial computationally, there are well-defined quantum mechanical methods for doing it, and AI is unlikely to be productive in this case.

However, after we have narrowed the field by removing problems which can be solved using conventional methods, there remain numerous problems in science waiting for AI solutions. It is important to scientists that solutions be found – science is all about understanding – but in our investigation of AI we need also to learn *why* the solutions are so difficult to find. Where do these complex problems which require AI come from? What 'causes' their complexity?

1.4 Why are there so many AI problems?

Combinatorial explosion

The game of chess is an excellent representation of many of the challenges facing AI; it illustrates a key reason why many problems are difficult to solve computationally.

The sequence of moves in a game of chess can be represented as nodes on a tree graph. The position at the start of the game is represented by the top node of the graph (Fig. 1.1) and all possible opening moves appear as nodes one level below this (the figure suggests just a few of these). Each node at this first level down corresponds to a different arrangement of pieces on the board after the initial move is completed. The second player can respond to any opening move in a number of ways; these options are shown as a further set of nodes at the second level (again, only a few are shown).

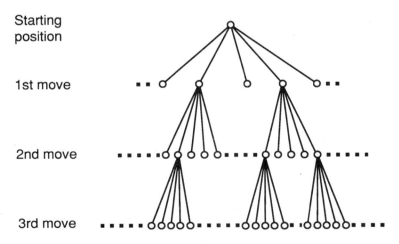

Fig. 1.1 A tree graph for a game of chess.

It is easy to see that the number of nodes on a graph of this sort grows extremely rapidly. Any attempt to look even half a dozen moves beyond the present state of the board may require the examination of millions of possible

Exhaustive search is a 'brute force' way of tackling problems, and is unlikely to be viable in any problem in which the number of possible solutions is very large.

An abbreviated exhaustive search in chess fails on two counts: it cannot foresee traps set by an opponent that will be sprung further ahead than the number of steps it assesses, and a program employing it uses no strategic planning.

sequences to find the best move. (The only sure route to success is to follow every conceivable sequence through to the end of the game; this is known as an **exhaustive search**, and in chess is an impossibly huge task. On the other hand, if we restrict our attention to a small number of moves ahead, what appears to be a wise decision now may, after a further four or five moves, turn out to be a foolish mistake, so an abbreviated exhaustive search is also not an effective strategy for a chess-playing computer that aspires to grandmaster status.)

This extremely rapid growth in the size of a problem is known as **combinatorial explosion**. A familiar chemical example is the growth in the number of isomers in a homologous series (Table 1.2).

Combinatorial explosion is responsible for the difficulties presented by many large-scale problems, both in science and in other fields. Nevertheless, you will recognize that, while a problem such as playing chess, or finding how many isomers of formula $C_{26}H_{54}O$ are ethers, becomes more difficult to solve as the number of nodes in the tree increases, *the problem itself becomes no more obscure*. The difficulty is not that the problem is in principle impossible to solve, but that the extent of the problem has grown very rapidly – *there are simply too many solutions*. This is a key discovery, and is a feature that characterizes numerous AI problems.

Among problems which display combinatorial explosion is a large class known as **NP problems** (non-deterministic polynomial problems), in which the difficulty finding the optimum solution by exhaustive search increases extremely rapidly with the size of the problem. The time taken to find a route through a maze depends exponentially upon the number of branches in the maze, and this is a typical NP problem. In Chapter 4, we shall encounter a chemical example: the chemical flowshop.

Table 1.2 Isomers of the alkanes C_nH_{2n+2}

n	Isomers
1	1
2	1
3	1
4	2
5	3
6	5
7	9
...	...
10	75
...	...
30	$> 4 \times 10^9$

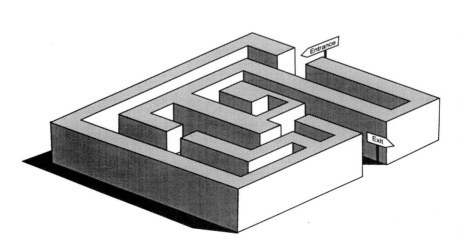

Fig. 1.2 A branching maze.

Problems of this level of complexity are widespread in science and elsewhere. Although their size generally precludes exhaustive search, AI methods offer a practical alternative in many cases.

It is important to realize that this alternative is not a short-term band-aid that we need just for the next few months or years. If you bear in mind the scale of these problems, it should be clear that AI methods do not represent a temporary way of attacking difficult problems, in the belief that, at some time in the future, faster computers will allow the use of conventional methods. These complex problems, by their very nature, cannot adequately be attacked by normal numerical methods, so *a solution to them using conventional algorithms is not brought noticeably closer by an increase in computer power, or by the use of more sophisticated statistical or arithmetic calculations.*

It is not extra computer power or further mathematical sophistication that is required, but a different approach entirely.

1.5 Understanding

Only humans solve crossword puzzles well. Computers generally use a modified search procedure in which any word that might have some tenuous relationship to the clue is checked to see if it will fit the crossword frame. Although this sort of search is just the kind of procedure at which computers excel, it is a particularly inappropriate method of solving a crossword, and the best computer programs fail miserably to beat even a mediocre human. The method that humans use cannot therefore simply be searching, or the computer would win every time.

The computer fails because it does not *understand* the crossword, and solving crosswords seems to be the kind of task that requires genuine understanding. The computer can search at high speed through the words that it knows to see which fit the frame, but it is incapable of making the intuitive jumps that characterize human reasoning.

Fortunately for science, AI programs can reason logically; in this way they can reproduce certain aspects of human thinking. They have as yet only a rudimentary ability to make what might be called intuitive jumps, (so are still not very good at crossword puzzles), but they can mimic many of the logical deductions of the brain. In solving scientific problems, AI programs must act intelligently, but the computer does not need to try its hand at lateral thinking. Understanding of the sort needed to solve crosswords is not required, so even AI programs of quite modest ability outperform conventional programs in a variety of scientific applications.

Ehrenstein's illusion (Fig. 1.3) illustrates the difference between human and machine understanding. Humans immediately notice the inner circle formed by the straight lines in this figure, even though the circle is not 'real'. Unless a computer were asked specifically to investigate what shapes are described by the end points of the lines, a conventional program would not discover the circle. However, image-analysis programs, which increasingly

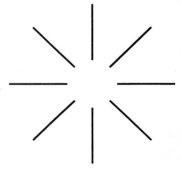

Fig. 1.3 Ehrenstein's illusion.

Humans notice Ehrenstein's circle in the figure above because it is brighter than its immediate surroundings.

incorporate AI elements such as neural networks, are adept at recognizing just such illusory figures and would find the phantom circle, without being explicitly told to search for it.

The Turing test

Computers do not understand much yet; their intelligence is at about the level of the ant. However, computer reasoning is developing rapidly, and many people accept that within perhaps a couple of decades a computer will be developed which can function intellectually as effectively as a person. It will then be hard to tell computer and human apart.

It is this thought which is behind an annual contest in America. Its purpose is to test whether a human can be fooled into thinking that the other half of a conversation he is having through the medium of a computer terminal is being provided by a second human, not a computer. A distinguished British mathematician, Alan Turing, who formulated many of the principles of modern computing, proposed a test, known now as the Turing test, on which the contest is based. The test involves what was at the time it was proposed a mythical entity – a computer and program of such sophistication that a human, conversing with it via a terminal, would be unable to tell whether another human or a machine was providing the other half of the conversation.

Surprisingly perhaps, some computer programs have already passed the Turing test in limited subject areas, such as 'Shakespeare's Plays', or 'Romantic Experience' (but not yet in science) and thus fooled people into believing that the programs are humans, not just intelligent software. It is a philosophical rather than a scientific question whether such a computer program can be said to *understand* the domain in which it works, rather than merely reason logically and in a convincing fashion about it. The difference between understanding and reasoning lies at the heart of this question, and is one which provokes enthusiastic debate among workers in the fields of AI and philosophy.

Outside the confines of philosophy, this relationship between reasoning and understanding, though fascinating, is largely academic. For the foreseeable future, computers and humans will think and understand in different ways. This difference should not concern us unduly. What matters to us is that human and electronic thinkers agree.

At one stage during the Second World War, 10 000 people in Britain worked on the decoding of German military messages; Alan Turing was one of them. He was among a group of linguists, mathematicians, and scientists working in British Military Intelligence at Bletchley Park, mid-way between Oxford and Cambridge. Turing worked on the 'Colossus' calculator and the famous 'Enigma' coding machines, two projects which were to prove crucial to the Allied war effort. Turing, who was a Fellow at Cambridge University, never lived to see his seminal ideas about computing put into practice; he took his own life in 1954 at the age of 42.

There is no evidence that the methods humans use to think are the 'best' available; they may soon be improved upon by computers. In turn, this may lead to still more sophisticated AI methods, thought up by the computers themselves.

1.6 Learning

Speech recognition and the motor control required to play tennis are the sort of tasks that humans manage well. Both require learning and memorization (though you should appreciate that 'learning' does not necessarily imply 'understanding'; we learn to recognize speech without having the least idea how we do it). If AI programs are to emulate humans, we can expect them to show this desire to learn and remember.

Learning is the improvement of performance through experience.

Just like humans, AI programs start life knowing nothing. They can manipulate data – but have no data to manipulate. To be of use, they need to gather information, and can do this in various ways:

How AI programs learn

- by a two-way conversation with the user;
- by being told what is true;
- by being shown examples of what they are required to interpret, such as mass spectral fragmentation patterns or infrared spectra, and learning to recognize characteristic features;
- by actively investigating their environment and learning from what they find there, using trial and error, or intelligent observation;
- by logical reasoning from what they already know.

Expert systems may use any of these methods to learn. Artificial neural networks and genetic algorithms usually learn by inspecting examples, or by investigating the environment.

This is quite different from the way in which a conventional computer program acquires information. Everything that such a program needs is built in, or is supplied as input data by the user. By contrast, AI programs contain an evolving, growing memory, which becomes progressively more valuable as the program becomes more experienced.

This memory gives the AI program advantages over conventional programs. A scientific AI problem can only rarely be reduced to a set of numerical calculations, whose execution leads unambiguously to the correct answer. Even if the problem can be formulated in this way, the time required to reach that solution is usually prohibitively long, unless we permit the program to learn. The numerical programs that are typical of most conventional scientific calculations are in AI replaced by programs that learn; indeed *learning is central to every AI method.*

As learning is critical to all AI methods, the area is often referred to as 'Machine Learning'.

1.7 Uncertainty

Just as there are problems which have millions of solutions, there are others that may not have even one about which we can be certain. Every day we meet questions to which the answer is a matter of opinion: 'Is heterocyclic chemistry incomprehensible?', 'Is baseball even more boring than cricket?', 'Are bald men sexy?'

If AI is to emulate human thought, it must be able to tolerate uncertainty of this sort, perhaps even use it to its advantage. It must deal effectively with questions that may have no definite answer, in the sense of there being none that are universally recognized as correct.

The genetic algorithm, which we shall meet in Chapter 4, is driven throughout by random numbers, so we might expect there to be some uncertainty in the answers that it provides. You may think it odd that a method which is completely dependent upon random numbers can provide answers that can be trusted at all. We shall see in Chapter 4 how this apparent contradiction is resolved.

All the methods discussed in this book are uncertain in the sense discussed here. Most expert systems have uncertainty built into the knowledge base, neural networks converge to a state which may depend upon the way they have been trained, and genetic algorithms rely heavily upon random numbers in their operation.

Uncertainty is often present in both the questions we ask, and the answers AI provides. Usually, the uncertainty does not arise from some fragility of the calculation, but because the problem itself can only be framed in imprecise form or because an unambiguous answer is not possible. Alternatively the AI method itself may unavoidably contain some random factor.

Uncertainty affects the behaviour of AI programs in two crucial respects:

● *The order in which an AI computer program executes its instructions may be unpredictable.*

If the same problem is presented in two different forms, an AI program may pursue two entirely different routes through its instructions to reach an answer. Some AI programs may even take a different route to the solution when the same problem is presented twice in identical form.

This will seem curious if you are familiar with normal computer programs, which always process the same instructions in the same order and reach the same conclusion, no matter how often they are run. However, most AI methods are themselves inherently uncertain. They may choose among several alternative courses of action, without an unambiguous rule to guide their choice. Or the way the program works may depend on how the problem is presented, or the order in which information about the problem is provided.

You might wonder how such programs could surpass the performance of a conventional program, when they are apparently so unpredictable. However, as we shall see, this element of uncertainty is deeply embedded in AI, and, paradoxically, is often crucial to its success.

● *The solution to a problem proposed by an AI method may not be unique; instead, several answers may be offered, each with a certain level of confidence.*

If an AI program tries to answer the question 'Are bald men sexy?' only those of us with little hair might believe an unambiguous 'yes'. The answers to some questions must be uncertain.

This introduces a second fundamental difference between AI and conventional programs. The latter usually reach a single (final) conclusion; an AI program, reflecting the less deterministic nature of its thinking, may not be absolutely certain that its preferred answer is correct, and so may propose this answer with some measure of confidence, or present a list of answers in order of preference.

This looks a bit weak-kneed. Science is largely about rules and certainties. If, after all its work, the best an AI program can do is to say:

Here is my answer. Well, actually you'll notice that there are several, even though you wanted only one, and I'm not exactly sure they are right, but there we are...you can't have everything

it may seem we are no further forward. We might even be more confused than when we started. It is a bit discouraging.

Recall, though, that the role of AI is to tackle the difficult, large-scale problems. As we have already suggested, these are usually very demanding computationally. They may be so huge that no known method can *guarantee* to deliver the right answer. If such problems cannot be solved exactly by *any* method in reasonable time, it is unrealistic to expect an AI algorithm to find rapidly *and be sure of* the exact solution. The best we can hope for is that an appropriate algorithm will yield an answer superior to any that could be reached using conventional methods, even if it is not guaranteed to be correct. AI methods might be better than conventional methods, but they do not have magical powers.

Uncertainty in the answer is therefore sometimes inevitable, and it is useful if the program can estimate how good the answer may be. Any answer is better than none; an estimate of its quality is bound to be valuable.

But this all-pervasive air of uncertainty is still rather unsettling. First we find it in the way an AI program works, and now it appears in the answers AI offers. Even if problems exist that are suited to attack by AI, it seems as though the success of any computation might be determined purely by chance: if the right path through the instructions that constitute an AI program happens to be selected, the calculation succeeds, otherwise it fails.

This is a rather pessimistic view, but there is an element of truth in it. The operation of many AI programs cannot be predicted with certainty. Yet this only mirrors our own experience and is not as undesirable as it may seem. We do not pretend that a question like 'Is baseball more boring than cricket?' cannot be posed, on the grounds that there is no 'right' answer. Uncertainty is a ubiquitous feature of everyday life and of numerous scientific problems also. If we choose AI to tackle vaguely-defined problems, the fact that our method may in some way be unpredictable does not prejudice our chances of success.

Thus, while AI may not be completely sure of an answer, its ability to deal with uncertainty endows it with considerable flexibility, so that most AI methods can be applied to problems in which the data themselves are uncertain. Each of the methods discussed in this book is capable of dealing with ill-defined or error-loaded data, and artificial neural networks allow us to make use of a method of growing importance in science, known as fuzzy logic.

Though uncertainty is woven inextricably into AI methods, this does not mean that the predictions of AI are less trustworthy than those of conventional methods. We must judge AI by its success in tackling real problems, and not by any preconceptions about the effect of uncertainty within the methods themselves. It is to the first of these methods that we turn in the next chapter.

It is not uncommon for problems solved using the genetic algorithm to have 10^{30} or more different solutions.

Fuzzy logic is designed to provide answers to such apparently unanswerable questions as: 'How big is a big molecule?', or 'How long is a piece of string?'

1.8 Applications of artificial intelligence in science

AI has been widely applied at the interface of science with other subjects. For example, expert systems are used in environmental planning, and genetic algorithms in assessing what characteristics make a face attractive. There have even been attempts (of limited success) to find the ideal muesli using genetic algorithms.

Chemistry and the other physical and life sciences offer AI problems which are remarkable in their diversity. In this primer there is space to touch upon only a few, but it is worth mentioning some areas in which AI is already used, to illustrate the breadth of application in what is still an emerging field.

AI has been applied to automated chemical synthesis, jet engine design, providing assistance to scientists in space, the human genome project, drug design, control of chemical flowshops producing drugs and fine chemicals, automatic feature recognition, image analysis, selection of optimum parameters for instrumental analysis, intelligent control of instruments, dispersal studies of pollutants, gas-phase kinetics, protein stereochemistry, querying of databases, studies of the origins of life, the design of synthetic routes in organic chemistry, and numerous other scientific tasks.

AI methods work. As a consequence, the range of applications is growing rapidly. In the 1970s and 1980s conventional programming was behind almost every significant application of computers to science, while AI methods were hardly used. In the coming years the situation will be reversed, as scientific programming comes increasingly to be dominated by AI.

We move on now to investigate how AI methods work, and look in more detail at the methods of greatest significance to chemistry and the physical sciences.

Further reading

Klir, G.J. and Folger, T.A. (1988). *Fuzzy sets, uncertainty and information.* Prentice-Hall, London.

Lauriere, J.-L. (1990). *Problem solving and artificial intelligence.* Prentice Hall, London.

Michie, D. (1986). *Machine intelligence.* Ellis Horwood, Chichester.

Shirai, Y. and Tsujii, J-i. (1984). *Artificial intelligence.* John Wiley and Sons, Chichester.

Simons, G.L. (1984). *Introducing artificial intelligence.* NCC Publications, Manchester.

Winston, P.H. (1984). *Artificial intelligence.* Addison-Wesley, Reading, Mass.

2 Artificial neural networks

2.1 Introduction

Computers are not much good at tasks at which humans excel, such as vision, speech recognition, and motor control. This difference cannot be due directly to a lack of speed, since a computer manipulates data thousands of times faster than neurons in the brain.

However, computer processors have a structure which is very different from that of the human brain, and it is natural to wonder whether human superiority in these areas might be related to the difference. Could a computer whose internal operations resembled those of the human brain really 'think'? And if it could, would it be as versatile as the brain?

This kind of speculation has led to the development of computer programs that copy what are presumed to be the workings of the brain, in the hope that these programs might show human-like intelligence, at least at a rudimentary level. The programs learn by training and experience, just as humans do, instead of following predetermined rules.

But there are formidable obstacles to be overcome before a computer can think like a person. One of the most fundamental is that our knowledge of how the brain works is fragmentary, so making a copy of it in silicon presents grave difficulties. Our understanding is still so incomplete that we are unable even to fully explain how the brain stores and recalls a simple fact, such as which day of the week it is.

Despite the difficulty of reproducing something that is only partly understood, great efforts have been made to mimic the behaviour of the brain using computers; this work has given rise to the field of artificial neural networks. Artificial neural networks have a long history, almost as long as that of the computer itself. Yet, at one stage, work in this area was virtually abandoned because it appeared that neural networks would be unable to solve all but the most trivial of problems.

An artificial neural network is an example of a **connectionist model**, in which many small logical units are connected in a network.

In the latter part of the 1980s a way around this difficulty was discovered, and since then neural networks have had some striking successes. Indeed there is now the extraordinary prospect that such networks might – unaided – discover new scientific laws. In this chapter we shall see how this prospect may become reality.

Artificial neural networks are computer programs based on a simplified model of the brain; they do not attempt to copy the fine detail of how the brain works, but try to reproduce its logical operation using a collection of neuron-like entities to perform processing.

2.2 Structure and operation of the brain

Since the design of artificial neural networks is inspired by the structure of the brain, we shall start by briefly reviewing the way in which the brain is believed to function.

The **neuron** is the fundamental processing unit in the brain (Fig. 2.1); a typical human brain has about 10^{10} neurons. Each one is linked to approximately 10^4 others by dendrites, which serve to deliver messages to the neuron. Each neuron also has an output channel known as an axon to funnel messages away.

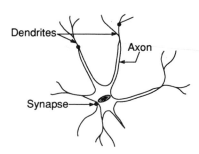

Fig. 2.1 A simplified model of a neuron.

The detailed biochemistry occurring at the molecular level in neurons is not fully understood, but the basic outline is clear. Signals are constantly flowing into neurons through the dendrites; these signals are processed by the neuron in a fashion which resembles the workings of a tiny logic device. Signals channelled into the neuron over a short period of time are integrated or summed biochemically. If the sum of these signals reaches a certain level, the neuron assumes an 'excited' state, whereupon it sends a signal out down its axon. Because of the highly-branched structure of the dendrites, a single outgoing message from the neuron is divided up many times and is transmitted to numerous other neurons.

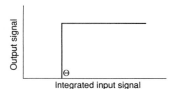

Fig. 2.2 The Heaviside threshold function.

The neuron behaves as a **threshold device**; it is quiescent unless the sum of the input signals over a period of time rises above some critical voltage, θ (Fig. 2.2). This voltage is determined by a **threshold function** which has a step-like shape, and is known as a **step function** or **Heaviside function**. If the threshold voltage is exceeded, the neuron is switched on and an output signal is generated.

Neurons that have assumed an excited state settle back rapidly into the quiescent state, but their message may by then have triggered into action many other neurons elsewhere in the network. Because of the richness of neural interconnection, signals propagate through the three-dimensional network of the brain in a manner that changes in an exceedingly complex way with time.

The network is constantly active as messages flicker from neuron to neuron, and as neurons respond to signals from muscles, or, via our senses, from the environment. This ceaseless activity is central to both thought and learning. Inter-neuron connections along a path that carries frequent messages are strengthened by the repeated stimulus and so, over time, the connections are reinforced. This preferential strengthening of well-used network links, which occurs as we master a new field or skill, appears to be the basis for learning.

It does not seem to matter much what the stimulus is: we can try to ski down a snow–field without falling, or learn the rules of mechanistic organic chemistry. Whatever the stimulus, an appropriate response can in principle be learnt, manifested in the strengthening of activated inter-neuron connections; when the right links are strengthened, we learn. It is exactly this mechanism which we seek to reproduce in an artificial neural network.

A crucial feature of the behaviour of the brain (and one which we can realistically hope to reproduce in an artificial network) is that *it need not be taught how to learn.* If a stimulus is presented repeatedly to the brain, it will learn to respond appropriately. It may be that evolution has wired a learning strategy into the brain for us in advance. Whether or not this is so, learning does appear to be essentially automatic, and seems to occur because of the way the brain is constructed. If this interpretation is correct, a computer program designed to reflect the structure of the brain should be able to learn in a human-like fashion.

Two important conclusions follow from this observation:

- It is not necessary to explain to an artificial neural network how to solve a problem.

- Artificial neural network programs are multi-purpose; with suitable training, a *single program* could solve problems in spectral interpretation, propositional logic, image analysis or fingerprint interpretation.

If you are familiar with conventional computer programs, you may be struck at this point by the contrast between what an ordinary program can do, and what, by contrast, we are suggesting a neural network can manage. The former can accomplish only the tasks for which it is specifically designed, while a neural network is a kind of generalized learning machine which can, in principle, learn almost anything. The potential of artificial networks in science, if they can be developed to a level of intelligence approaching that of the human brain, is therefore enormous.

This is related to the 'built-in' knowledge called instinct. When they leave the egg, nest-building birds already have the knowledge needed to build nests, as can be demonstrated by removing eggs from the nest and rearing the chicks that hatch from them in isolation from other birds. Evidently nest-building is 'hard-wired' into the brains of these birds. It is unclear, though intriguing to speculate about, what knowledge might be hard-wired into the human brain at birth.

Note the 'or' in the second of these points. A given program could be trained to accomplish only one of these tasks; it would not be able to interpret both spectra and fingerprints, for example.

2.3 The elementary perceptron

Chemically, a neuron is an intricate and subtle device, relying on the behaviour of ions in solution. It can detect and sum minute signals, assess the size of that sum, and transmit messages to other neurons.

Although the biochemistry within a neuron is quite complex, the logical operations that it performs are by contrast rather simple:

The functions of a simple neuron

- A neuron has multiple input connections, and in some way can add up signals arriving on these connections.
- If the sum of the inputs is below a threshold value, the neuron is quiescent and remains 'off'.
- If the sum reaches the threshold level, the neuron is turned 'on' and a message is sent out.
- The neuron returns to a quiescent state after a short time.

The elementary perceptron is an example of a **feedforward system**; incoming signals pass through it in the forward direction only to provide an output.

It is a straightforward matter computationally to make an artificial neuron with these characteristics; this is known as an **elementary perceptron** (Fig. 2.3). The perceptron is a very simple feedforward system, and a crude approximation to a neuron, but with a small modification it will form the building-block from which our artificial networks will be constructed. Even on its own it can learn to perform some useful tasks, and an understanding of how it does this will help us to appreciate later how a network of perceptrons can learn.

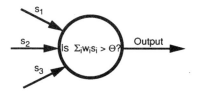

Fig. 2.3 An elementary perceptron

How a perceptron learns

The perceptron is a decision-making unit with several **input connections** (also known as **synapses** by analogy with the neuron) and a single **output connection**. A signal s_i which is delivered from input i is multiplied on arrival by a **connection weight** w_i, so that each signal appears at the perceptron as the weighted value $w_i s_i$. The perceptron sums the incoming signals to give a total signal $S = \Sigma_i w_i s_i$.

The perceptron applies a Heaviside threshold function, with threshold θ, to the sum S; if the sum reaches the threshold level, the perceptron fires; if the sum falls below this level, it remains quiescent.

$$\textbf{Input: } \sum_i w_i s_i \qquad \textbf{Output: } = 1 \ if \ \sum_i w_i s_i \geq \theta, \quad = 0 \ if \ \sum_i w_i s_i < \theta$$

The connection weights to a perceptron can be thought of as amplifiers, whose purpose is to adjust the size of a signal before forwarding it to the perceptron.

The behaviour of a perceptron is determined by the weights of input connections to it and by the level at which the threshold is set. Knowledge is stored as the values of these adjustable parameters, so a novice perceptron which knows nothing begins life with all its connection weights set to random values. Learning is then the process of adjusting the weights in a way that roughly parallels the training of a biological system. Through a gradual modification of the weights, the perceptron can learn to perform a variety of computations, including elementary image recognition.

This learning follows much the same course in the perceptron that it might in a dog or a child. Just as a dog may be rewarded with praise or treats for good behaviour, and punished for bad, the perceptron is trained by being given input data and asked to make a decision about its nature. If the conclusion the perceptron reaches about the data is incorrect, it is punished; if it makes the right choice, it is rewarded, in each case through the modification of connection weights. As this training process is repeated, the perceptron gradually begins to differentiate between desirable and undesirable behaviour.

We will illustrate perceptron learning by considering how it might solve a simple chemical problem, that of distinguishing between the structural formulae of molecules containing a cyclohexane ring, and those without one (Fig. 2.4). This is an example of a problem within a field of central importance in AI – that of image recognition – and is an area in which artificial neural networks show particular promise.

The procedure described here is known as **supervised learning**, since we know before starting what we are trying to achieve.

We should not press the analogy with human learning too far. The way a human learns suggests how we *might* teach an artificial network, but does not provide a prescription that we *must* follow.

Training

Any artificial neural network, including the perceptron, starts from a position of complete ignorance, so a training period is required before it can be let loose on real problems. During training, the novice perceptron is shown examples of what it must learn to interpret; these examples constitute the **training set**. This consists of two parts: the **training stimulus** is a collection of inputs to the perceptron – in this instance the structural formulae of molecules in the training set. Associated with each training stimulus is a **training target**, which is the desired output (the 'right answer') for each stimulus – in this example each target is the classification of the corresponding formula as a ring-containing or ring-absent structure.

Initially the perceptron knows nothing of the difference between cyclic and straight-chain molecules; indeed, it is unaware even of the nature of the task which it faces. If it is to learn to recognize six-membered rings in a molecule, a mechanism must be provided by which input signals can convey to it structural information about each molecule within the training set. We can achieve this by taking the inputs from a bank of small light-detecting sensors, each of which views a portion of a video screen on which formulae can be displayed (Fig. 2.5).

A structure from the training set is shown, and the perceptron allowed to make a decision on its identity. Its decision is compared with the training target and the connection weights are adjusted in the manner described below. After this adjustment, another structure is shown and the process is repeated until the perceptron reaches the required level of proficiency in identifying structures.

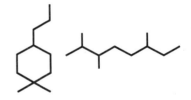

Fig. 2.4 The perceptron must learn to distinguish between these molecules.

The formulae are shown in a standard orientation, in which any ring always appears in the same region of the screen, so that the perceptron is not misled into believing that the same formula displayed in two different orientations is a different molecule.

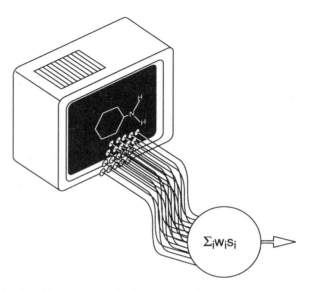

Fig. 2.5 How structural information is fed into the perceptron.

When a structure is shown, each sensor sends a signal determined by the amount of light it sees: a strong signal will be output by sensors viewing white atoms or bonds, a weak signal from those viewing parts of the screen displaying dark space (Fig. 2.6).

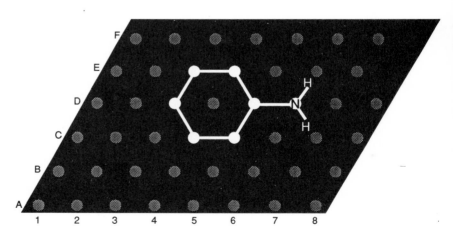

Fig. 2.6 What the sensors see.

The signals from the sensors, multiplied by the appropriate connection weights, are passed to the perceptron, which computes their sum and checks to see whether this reaches the threshold. If it does, the perceptron outputs an 'on' message; we will take this as an indication that the perceptron thinks it has recognized a cyclic molecule. If the weighted sum of the input signals falls below the threshold, the perceptron will remain 'off', and we will take this to be its indication that it thinks it has found a molecule without a six-membered ring.

The output of the perceptron is compared with the training target for the molecule on the screen and the perceptron is punished or rewarded accordingly. If it has made the right choice, the connection weights are left unchanged; the meagre 'reward' the perceptron receives is that it is not punished. If, however, the perceptron has made the wrong choice, the connection weights must be altered so that it is less likely to make the same mistake in the future.

Suppose the perceptron fired when it should not have; the sum of the weighted signals reaching it must have been too great (otherwise, it would not have fired), so connection weights on all *active* inputs are lowered. Weights of connections to sensors which sent no signal are unchanged; there would be nothing to be gained by altering these since, if a sensor is inactive, no change in weights of connections to it will alter the signal arriving at the perceptron, and if changes were made, any knowledge built up in the weights by previous training might be damaged.

If the perceptron remained inactive when it should have turned on, the total signal that the perceptron sees and integrates needs to be greater, so weights on active connections must be increased to encourage it to fire. Once again weights on inactive connections are left untouched.

We can summarize these learning rules as:

Elementary perceptron learning rules

- If the output of the perceptron is correct, do nothing.
- If the output of the perceptron is 'on', but should be 'off', decrease the weights on active inputs.
- If the output of the perceptron is 'off', but should be 'on', increase the weights on active inputs.

The choice of how output from the perceptron is interpreted is purely arbitrary. If we had chosen to take 'on' as an indication of the *absence* of a ring, training would lead to the development of different connection weights, but the perceptron would eventually be no less proficient at distinguishing one type of molecule from the other.

Once the weights have been adjusted (if necessary), another structure is chosen from the training set and the process is repeated.

What will be the effect of this training? Because structures are displayed in a standard orientation, some sensors will usually see dark space when one class of molecule is displayed, but see a light bond or an atom when a molecule in the other class is shown. For example, sensors C4, C5, D3, D5, E3, and E4 in Fig. 2.6 will each see an atom when a cyclic molecule is shown, but most of them will see dark space when a non-cyclic molecule appears.

Each time a non-cyclic molecule is shown, the perceptron should remain off; if it mistakenly turns on, the connection weights on all active inputs will be decreased according to the learning rules. Conversely, each time a cyclic molecule appears, connection weights to those sensors viewing light bonds or atoms will be increased by the learning rules, if the perceptron has not noticed the ring.

As successive structures are shown in this **training mode**, the perceptron will gradually scale the connection weights in a manner that improves its ability to discriminate between the two types of structure. Eventually, when a straight-chain molecule is shown and most of the group of sensors identified above see dark space, a small signal will always reach the perceptron; it will remain 'off', and therefore output a 'straight-chain molecule' message. If a cyclic molecule is shown, the crucial sensors over the ring see light atoms and will transmit a large signal which will persuade the perceptron to turn on and send a 'cyclic molecule' message. The perceptron has learnt to solve the problem, and it has done so without us having to explain to it how this should be done.

The learning rules for the perceptron are easy to implement, but imprecise. They define *which* weights should be adjusted, but they say nothing about the *size* of the adjustment. One recipe we could use is to raise or lower the weight on every active connection by a fixed proportion of its current value, say 10 per cent, every time modification is needed. This should pull the connection weights towards a solution, but it is limited in its approach. If the output from the perceptron is wrong, but only just (the perceptron nearly managed to fire, perhaps, but the total signal fell just short of the threshold), a minor change to connection weights would be appropriate, but if the perceptron was hopelessly wrong (the total signal was quite different from what was needed to give agreement with the training target), a larger change would be desirable. This can be taken into account by adjusting the weights by an amount proportional to the **error signal**, that is, the difference between the target output and the actual output, and this usually improves the rate of convergence towards a solution.

As we shall see later, the error signal plays a crucial role in the training of a full neural network.

The training set

Each complete pass through the examples in a training set is known as an **epoch**.

It is not possible to predict how much training will be needed before the perceptron becomes proficient at identifying structures. Though perceptrons are fast learners, many passes through the training set will be needed.

The structures shown to the perceptron will cover a representative range of the sort of molecules that it must learn to recognize, but the training set need not include every possible molecule. It is the purpose of training to provide examples from which the perceptron can extract the crucial features that distinguish one class of molecule from the other. The examples must therefore be representative, but need not be all-encompassing.

Neural networks can be trained to distinguish photographs of crystalline and amorphous material without being shown every possible type of crystal, because networks can learn the distinguishing feature – the presence or absence of straight edges – from a comparatively small group of samples.

Once training is over, the connection weights are fixed to their trained values. Performance of the perceptron can be tested by showing it an unknown structure and asking for an identification. In this **production mode** the perceptron defines an input $\rightarrow$ output mapping; a given structural formula on the screen always triggers the same predictable conclusion about the class to which the structure belongs.

Provided testing shows that training is complete, the perceptron should now be able to categorize every member of the training set. However, if this were all it could do, its achievement would be modest. Fortunately, its ability extends further; it will be able to identify structures which it has never seen before, but which bear some direct relation to members of the training set. New images will be recognized provided they lie within the confines of what it has learnt about, but it is not possible to go beyond the logical information contained within the training set. While the perceptron will recognize the presence of a six-membered ring in molecules, it does not have the knowledge needed to go beyond these limits, so it could not, for example, consistently identify molecules containing pentane or heptane rings. Put simply, it can interpolate, but not extrapolate.

Though we cannot expect the perceptron to recognize a pentane ring if it has been brought up on a diet of six-membered rings, it has nevertheless developed a significant skill: it has discovered how to translate visual information into logical information. This is a skill of very widespread application. The interpretation of photographs from spy satellites, microscope pictures of stained cells, or the application of computer vision each rely upon logical deductions derived from visual clues. Of course the elementary perceptron is a rather unsophisticated entity and far too simple to find a place in the vision system of a robot. In fact, as we are about to see, it suffers from a fatal flaw. Nevertheless, that such a simple entity can successfully turn visual information into logical information gives an early hint of the power of a full artificial neural network.

The perceptron is effective for certain tasks such as the molecular structure classification problem, in which it can learn to reliably distinguish between isomers. Not every problem in chemistry is this straightforward however, and we shall describe now the difficulty which brought research in artificial neural networks to a virtual standstill for nearly two decades.

2.4 Linearly inseparable problems

Perceptrons are fast but simple creatures. The type of problem that they can solve is very restricted: the molecular structure problem, for example, is soluble only if the structures are always shown with the same size and orientation on the screen. If these requirements are relaxed, a given sensor will see dark space for some orientations of a molecule, and a bond or an atom when the same molecule is shown at a different angle. It is not surprising that the perceptron then becomes confused and cannot learn.

We can illustrate the very general nature of the difficulty facing the perceptron, and discover what this difficulty is, by considering a second chemical application: the monitoring of gas-phase samples for the presence of pollutants.

Long path-length infrared absorption spectroscopy is a suitable analytical technique for identifying samples in the gas-phase. A gaseous sample is held within a cell which has reflecting mirrors at both ends. The mirrors reflect the

The inability to extrapolate is an important limitation to neural network models. British Rail have carried out research on an artificial neural network whose job would be to watch unmanned level crossings for people or vehicles on the line when trains are approaching. Such a network might eventually be successfully trained to recognize people and cars by giving it the chance to observe many examples of such things on the crossing, but training it to interpret less common images, such as fallen branches, a newspaper blowing in the wind, or a stray cow is not trivial. The consequences of a neural network working incorrectly in such a situation may be serious, not least for the cow.

Although the perceptron learns to solve the problem, it does not of course 'understand' what it has done; the algorithm has simply evolved connection weights that permit it to identify structures with an acceptable success rate. It is, however, no less useful because it does this without a human-like understanding.

infrared beam repeatedly, to give an effective path length many times the physical length of the cell, and thus increase the total infrared absorption.

Suppose samples taken close to two industrial sources are to be assessed on a routine basis. Each sample is mainly air, but occasional samples may show the presence of a single pollutant which may be either ethylamine (Fig. 2.7(a)) or nitrous oxide (Fig. 2.7(b)).

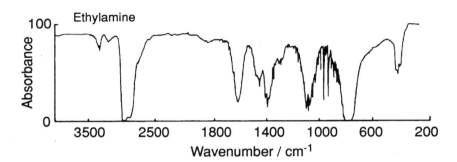

Fig. 2.7(a) The infrared spectrum of ethylamine.

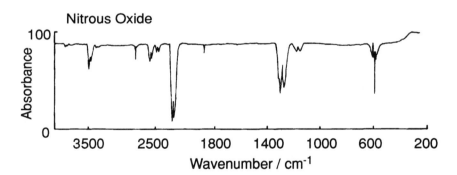

Fig 2.7(b) The infrared spectrum of nitrous oxide.

Because many samples must be analysed, the analyst decides to automate the process and use an elementary perceptron to look through each spectrum as it is recorded to determine whether a pollutant is present.

The training set consists of a number of spectra, some of which show the presence of ethylamine or nitrous oxide. As the perceptron can work only with a finite number of input signals, each spectrum is digitized, and passed to the perceptron as a list of absorbance measurements at certain fixed wavelengths; these absorbance measurements comprise the input signals (Table 2.1).

After many passes through the training set, the perceptron in production mode is left to monitor spectra unattended and use its knowledge to watch out for pollutants; if one or other is detected it can alert the analyst.

This is a two-state classification task not unlike the six-membered ring problem (pollutant present/not present), but there is a hidden difficulty. In operation the perceptron functions reliably until it encounters the spectrum shown in Fig. 2.8. Although the spectrum is clearly different from that of either pollutant, the perceptron incorrectly reports finding one. What has gone wrong?

Table 2.1 Part of the digitized infrared spectra of nitrous oxide and ethylamine.

Wavelength	% Transmittance	
cm^{-1}	NO	C$_2$H$_5$NH$_2$
600	60	76
700	88	54
800	89	0
900	89	52
1000	90	77
1100	91	18
1200	81	80
1300	48	72
1400	90	16
1500	91	50

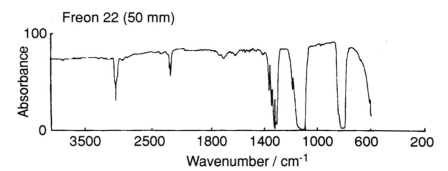

Fig. 2.8 The infrared spectrum of Freon 22.

To understand the difficulty, we note (Table 2.1) that the spectra of the pollutants show significant absorbance (that is, low transmittance) at either 1300 cm^{-1} or at 800 cm^{-1} (among other wavelengths). During training the perceptron will learn that high absorbance near 800 cm^{-1} or 1300 cm^{-1} indicates the presence of a pollutant. However, what it does not know is that if peaks appear at *both* positions this is an indication of the *absence* of both pollutants (since any sample contains at most one of them).

The spectrum with which the perceptron cannot cope is that of Freon 22. This halogenated hydrocarbon is implicated in the destruction of the ozone layer in the atmosphere, and therefore is a pollutant in its own right, but the perceptron does not know that, and its job is not to signal the presence of *any* pollutant, but only the two that the analyst is interested in.

The perceptron has been fooled by the fact that absorption in the Freon spectrum is high just where it is expected: at 800 cm^{-1} and at 1300 cm^{-1}. Indeed, since the perceptron finds two peaks that it is looking for in the 600–1500 cm^{-1} region, we can imagine that it might feel especially confident in its diagnosis. This problem is a variant of the **XOR problem**; XOR(*x,y*) is true if exactly 1 of *x* and *y* is true (Table 2.2).

Table 2.2 The XOR problem

	XOR Problem		IR Problem		
x	y	XOR(x,y)	Peak at 800 cm^{-1}	Peak at 1300 cm^{-1}	Pollutant present
0	0	0	no	no	no
1	0	1	yes	no	yes
0	1	1	no	yes	yes
1	1	0	yes	yes	no

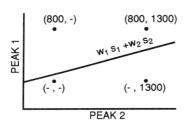

Fig. 2.9 Solutions to the infrared spectra problem.

The solution to a linearly-separable problem is a linearly-weighted sum of the inputs, and this is a computation within the capabilities of the perceptron.

The solutions to this problem are mapped in Fig. 2.9. Output from the perceptron is determined by how the weighted sum of inputs, $w_1s_1 + w_2s_2$, compares with the threshold θ. The sum $w_1s_1 + w_2s_2$ defines a straight line which is superimposed on the map of solutions to the problem in the figure. The perceptron must learn that a pollutant is present if a peak appears at either 800 cm^{-1} or 1300 cm^{-1}, but that no pollutant is present if peaks appear at neither or at both of these wavelengths. The straight line $w_1s_1 + w_2s_2$ must thus partition the points in Fig. 2.9 into two groups: one containing {(800,-) and (-,1300)}, and the other containing {(-,-) and (800, 1300)}. It is clear that no straight line can effect this partition, so the problem cannot be solved by the perceptron.

This problem is not **linearly-separable**. Every solution to a linearly-separable problem can be placed into one of two sets separated by a straight line. If a problem has a linearly-separable solution it can be shown that it is always possible for the perceptron to find it. As Fig. 2.9 suggests, the linearly-separable problem is the only type that the elementary perceptron can solve.

This is a crippling restriction. Most significant scientific problems are not linearly separable, and of the small number that are, almost all are better tackled by other methods. This is also true for problems outside science, and recognition of this difficulty, and the inability of the elementary perceptron to circumvent it, brought research in this area to an abrupt halt.

2.5 Multi-layer neural networks

The failure of the perceptron to handle real-world scientific problems highlights the rather tenuous resemblance between it and our model of the brain; there is little similarity between biological and artificial systems, and this difference might be one reason for the restricted capabilities of the latter.

Additional power is essential if anything resembling a neuron is to be useful in science.

At the start of this chapter we commented on the differences between the structure of the brain and that of the computer. It is worth looking at those differences a little further to provide clues about how the performance of the perceptron might be improved.

Most electronic computers are **von Neumann** or **serial** machines: data are widely dispersed in the computer's memory, but all processing takes place at a single point (Fig. 2.10). This unique processing unit forms a bottle-neck through which all instructions must be squeezed, and it is the speed at which this can be done which ultimately determines how rapidly the computer can operate.

By contrast, the brain can process several independent streams of information simultaneously: we might (a) be taking notes in a lecture, at the same time as we (b) are watching what the lecturer writes on the board, and (c) listening to an explanation of the lecture material, (d) all the while being vaguely aware of how uncomfortable the lecture room seats are. Each step requires separate, but simultaneous access to and use of the brain. If we want to write a few words, or read what is written on the board, we do not have to stop thinking to do it.

Just as the data that the brain processes are widely distributed within it, so also are the processing units. Indeed the distinction between a memory unit and a processor in the brain is blurred at best, since neurons form a part of both types of structure. Processing of different items of information occurs simultaneously, as many neurons work on different parts of the data, so the brain is referred to as a **parallel** device (Fig. 2.11).

In our desire to construct a program that can think, we have limited the chances of success by using just a single computational unit. One perceptron is unlikely to be able to accomplish much on its own, and the obvious difference between this and the rich network of neurons that makes up the brain suggests that a promising step would now be to add more perceptrons. This can be done in two ways: first by giving the perceptrons neighbours to form a layer of units which share input from the environment; and secondly by introducing further layers, each taking as their input, the output from the previous layer.

The first set of units in the resulting network, (Fig. 2.12), is an **input layer**, whose purpose is to distribute incoming signals to the next layer; it does no thresholding, so the units here are not perceptrons; they are sometimes known as 'sensor elements'. The perceptrons in layer 2 constitute a **hidden layer**; they can communicate with the environment only by receiving or sending messages to units in the layers to which they are connected. The **output layer** provides a link between the artificial network and the outside world.

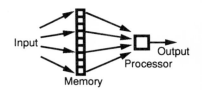

Fig. 2.10 Serial calculation.

It is convenient to be able to think and watch at the same time, but it would also be valuable to be able to *think several thoughts* simultaneously. We could concentrate on a lecture, and at the same time review quantum chemistry, plan an essay, and catch up with the maths course. Maybe you suspect you can do this already; if so, prove it by doing the following calculations *simultaneously* in your head: 273.15 × 3.1415926 and 981 × 1.4142. If you cannot do this, ask yourself why you can think and observe, or think and listen, but not think and think simultaneously. This is one area in which neural networks (on parallel computers) can already outperform their human counterparts.

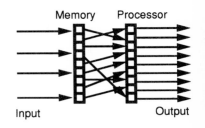

Fig. 2.11 Parallel calculation.

There may also be connections between perceptrons in non-neighbouring layers, or 'feedback' connections which permit signals from units in one layer to be returned to units in an earlier layer. Such **recursive networks** are less widely used than layered networks, but are valuable in certain specialized applications.

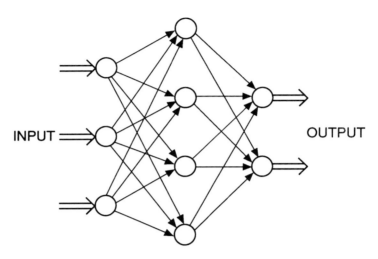

Fig. 2.12 A simple neural network.

Every perceptron in the type of network shown in Fig. 2.12 is connected to all units in adjoining layers, but there are no connections between units in the same layer. This is a **fully-connected layered feedforward network**. Since the network is feedforward, in production mode messages flow in the forward direction only.

The molecular structure and the infrared analysis problems have illustrated that learning is partly a process of categorization. By recognizing something, we assign it to a category about which we have some defining information. Grouping it in this way is equivalent to separating it from all other items which, though they may have some similar characteristics, belong in a different class. We have seen that an elementary perceptron can divide a set of items into two classes, provided that the items are linearly-separable (Fig. 2.13), and we will show now that a layered network can be used in applications for which the solutions are not linearly-separable.

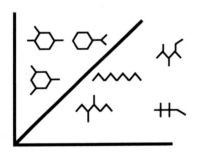

Fig. 2.13 A linearly-separable problem.

A linearly-inseparable chemical problem

Many problems in science are the conjunction of two or more linearly-separable problems. For example, consider how an artificial neural network might be used to monitor the temperature and pH of material in a reaction vessel (Fig. 2.14).

Let us suppose that the temperature inside the vessel must not rise above 95°C, nor the pH fall below 4.5; if either event occurs, the network must sound an alarm. One perceptron on its own cannot deal with this problem because there is no linearly-separable solution, but a small network can (Fig. 2.15).

Input to the hidden layer of this network is provided by pH and temperature probes, each of which provides an analogue (continuously-variable) signal determined by conditions within the reaction vessel. One perceptron in the hidden layer monitors the temperature: it turns on if the temperature exceeds 95°C, but ignores completely the signal from the pH probe. The second unit in the hidden layer monitors the pH, and turns on if the pH falls below 4.5, but ignores input from the temperature probe. Output from the two perceptrons in the hidden layer is combined at the single perceptron in the output layer; if either hidden unit sends it an 'on' signal, conditions in the vessel are outside the prescribed limits and the output perceptron is triggered to send a signal.

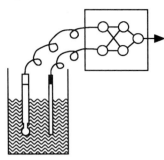

Fig. 2.14 A neural network monitoring conditions in a reaction vessel.

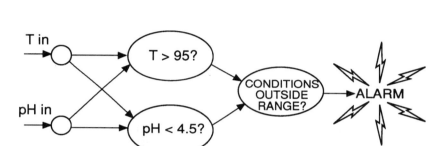

Fig. 2.15 A network that can solve the reaction conditions problem.

This simple network classifies conditions within the reactor according to two independent criteria, and therefore can be used in situations in which solutions are linearly-inseparable. Reactor conditions are mapped in Fig. 2.16; it is the role of the network to determine when those conditions fall outside the shaded area.

By adding further units we can restrict the shaded area to a closed figure (Fig. 2.17), known as a **convex hull**, which can be made of any shape desired, provided enough units are added to the network. A larger network could monitor the material in the vessel for various other physical parameters (for example viscosity, a lower bound to the temperature, etc.), and either warn the experimenter when conditions are inappropriate, or perhaps activate heaters or autotitrators to adjust conditions intelligently.

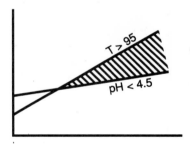

Fig. 2.16 A mapping of conditions in the reaction vessel.

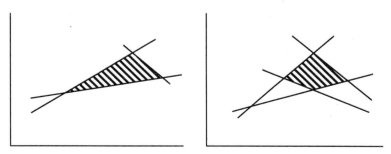

Fig. 2.17 Some convex hulls.

Similar networks can be used to solve the gas-phase pollutant problem (Fig. 2.18) .

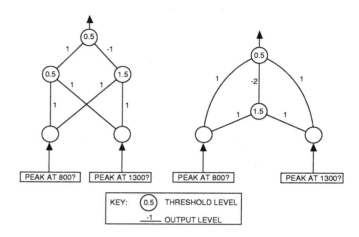

Fig. 2.18 Two networks that can solve the infrared classification problem.

This is beginning to seem rather encouraging. We have now a network which will do something useful, and do it reliably; furthermore this network obviously bears a much stronger resemblance than the perceptron did to the highly-interconnected structure of the brain, and this suggests that we are perhaps approaching something capable of genuine decision-making. It is therefore disappointing to discover that though a layered network of elementary perceptrons appears to have quite wide-ranging powers, it is actually no smarter than the perceptron itself.

The credit assignment problem

The difficulty is that though the perceptron can intelligently monitor our apparatus, *it cannot learn to do so*. By setting up a network manually (in other words, specifying the connection weights and threshold limits in advance, and omitting the training period), an artificial network can handle the pH/temperature monitoring problem. However, the network cannot learn suitable connection weights for itself – it must be told what they are.

This is not much use. Few applications are simple enough that suitable connection weights and thresholds can be determined by hand; in applications like the present one, it would be preferable to use electronics to monitor conditions within the reaction vessel. What is needed is a network that can learn for itself, not one that needs hand-holding by the experimenter.

The reason the network is unable to learn is because each perceptron in the hidden layer, which accepts input data, only passes the message 'on' or 'off' to the next layer, and this simple binary message effectively isolates the output layer from input to the network. An output perceptron is unable to determine whether a perceptron in the hidden layer to which it is connected has been turned on by the sum of its inputs only barely reaching the threshold level, or whether the threshold has been greatly exceeded. The perceptrons cannot fire timidly or enthusiastically – they either fire or do not.

The hidden layer has thrown information away. When this layer intervenes, it causes an insoluble problem in the assignment of weights connecting elementary perceptrons: if an output unit is active when it should not be, connection weights somewhere in the network must be adjusted to turn it off, but which weights? Should it be the weights of connections to the output unit from active units in the hidden layer? Those units are, after all, directly responsible for the output unit exceeding its threshold limit and firing. Or should weights be changed to prevent the hidden layer units themselves from firing?

Unfortunately, the network has no way of knowing which of these alternatives is right. *Training is meant to teach the network to respond to a given input with the appropriate output, but the network cannot learn to do this unless the output units have some information about what the inputs to the network are.* In this model, input signals are completely hidden from the output layer, and it is therefore impossible for the network to adjust the connection weights so that a certain input triggers the appropriate output.

This is the **credit assignment problem**, and it prevents a network of elementary perceptrons from learning how to solve linearly-inseparable problems. The complexity of the model has been increased by the addition of extra perceptrons; it is now more capable, yet it cannot learn, and this severely limits its usefulness.

Modified threshold functions

The perceptron network can perform some useful tasks, but we need to solve the trick of getting it to learn, so that the network can find suitable connection weights by itself. Fortunately, there is a straightforward way to accomplish this, and a network that can learn successfully can be built by borrowing a further idea from models of the brain.

A neuron is not a binary unit, limited to 'on' or 'off' in the way that our perceptron is; instead, it can respond with an output signal graded by the size of its input signals. We can reflect this in an artificial network by modifying the threshold function that perceptrons use, since it is the all-or-nothing nature of the step function that isolates the output layer from input to the system. If we combine a multi-layer structure with a more flexible threshold function, a network can be constructed that can solve complex problems with great ingenuity.

Two different types of threshold function are commonly used in feedforward layered networks. If the perceptron employs a **linear threshold function** (Fig. 2.19) the output of the perceptron is zero for integrated input

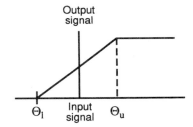

Fig. 2.19 A linear threshold function.

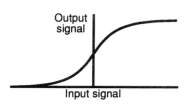

Fig. 2.20 The sigmoidal threshold function.

The sigmoidal function has the particular advantage that it provides a kind of automatic gain control. Large and small input signals may be presented simultaneously to the network, on different input lines. It is easy to see that information carried by small signals may then be overwhelmed by much larger signals. When a sigmoidal threshold function is used, large signals can still be accommodated by the network, and small signals are passed on with little attenuation, so do not become insignificant compared with their larger competitors. The sigmoid function also has a differential that can be simply expressed in terms of the function itself, which speeds computation.

levels below θ_l, and fully on for inputs above θ_u. Between these limits, the output signal is linearly related to the total input signal. An adapted perceptron of this sort that is only just turned on by its inputs will reflect this by transmitting a correspondingly small output signal. Clearly this threshold function allows some information about signal size to be passed from one layer to the next, which is necessary if the network is to learn.

The most widely-used threshold function is the **sigmoidal threshold function** (also known as the 'sigmoidal squashing function' or 'logistic function'). This smoothly varying function approaches asymptotically the extremes at which the perceptron is fully on or fully off (Fig. 2.20).

The sigmoidal threshold function is:

$$g(n) = (1 + e^{-n})^{-1} \tag{2.1}$$

which has limits of 0 at $-\infty$ and $+1$ at $+\infty$; n is a positive number which measures the 'spread' of the function; as $n \to \infty$, $g(n) \to$ the Heaviside function.

A multi-layer network which incorporates a sigmoidal threshold function is certainly more reminiscent of the structure of the brain than the elementary perceptron is, and we can realistically expect it to have much wider powers. Networks of this sort are now widely used, but because of their complexity, the rules by which these networks learn differ from those used for the elementary perceptron. The central difficulty is knowing how the credit assignment problem should be solved; how should weight changes be divided up between the various connections? It is to this learning problem that we turn next.

2.6 Backpropagation

In the elementary perceptron, learning is simple and the learning rules unambiguous. Matters are more complicated in a network, and we must establish how changes in connection weights should be allocated to connections between different layers to promote learning.

The most widely-used solution to this problem is **backpropagation** (literally dozens of other less popular paradigms exist). The error signal $(t_{pj} - o_{pj})$ is calculated, where o_{pj} is the actual output from output perceptron j for training set member p, and t_{pj} is the target output. A proportion of the error signal is allocated to the various connections in the network and the connection weights are adjusted according to a mathematical prescription whose objective, as in the elementary perceptron, is to reduce the error signal.

In **stochastic backpropagation**, connection weights are adjusted every time a member of the training set is shown to the network. In **standard backpropagation**, the error signals are collected for all output units and all training targets, and the connection weights are adjusted at the end of every epoch, that is, after all members of the training set have been shown to the network once.

The error function E_p is defined as (half of) the sum of the squares of the error signals for all units:

$$E_p = \frac{1}{2} \sum_j (t_{pj} - o_{pj})^2 \qquad (2.2)$$

The objective of backpropagation is to minimize E_p by **gradient descent**, an iterative least squares procedure in which the algorithm tries to adjust connection weights in a fashion which reduces the error most rapidly, by moving the state of the system downwards in the direction of maximum gradient.

The weight of a connection at stage $(t + 1)$ of the training is related to its weight at stage t by the equation

$$w_{ij}(t+1) = w_{ij}(t) + \eta \delta_{pj} o_{pj} \qquad (2.3)$$

in which η is a gain term, known as the **training rate factor**. It is possible to derive expressions prescribing the size of the changes that must be made in connection weights to reduce the error signal.

For the output layer: $\delta_{pj} = k o_{pj}(1 - o_{pj})(t_{pj} - o_{pj})$ (2.4)
For the hidden layer: $\delta_{pj} = k o_{pj}(1 - o_{pj}) \sum_k \delta_{pk} w_{jk}$ (2.5)

These expressions, which for historical reasons are often known as the **generalized delta rule**, show that the extent of the adjustment of connection weights to hidden layers depends upon errors in subsequent layers, so modifications are made first to the output layer weights (which is straightforward, since for this layer we know both the actual output and the target output), and the error is then propagated successively back through the hidden layers.

Each unit receives an amount of the error signal which is in proportion to its contribution to the output signal, and the connection weights are adjusted by an amount proportional to this error.

Backpropagation by gradient descent is a generally reliable procedure. Nevertheless, it has its limitations. First, it is not a fast training method, particularly if the surface defined by the error function is relatively flat (in other words, the size of the output error is not strongly dependent upon the connection weights). Changes in connection weights then do little to reduce the output error and convergence to ideal weights is slow.

Secondly, since gradient descent is analogous to rolling a ball down the surface defined by the error function to find the lowest point, it is possible for the network to reach a sub-optimal set of connection weights corresponding to a local minimum in the surface. Gradient descent will be unable to reduce the error further, as this requires first moving the connection weights towards a less favourable solution. This situation can sometimes be avoided by using a different algorithm for training.

When stochastic backpropagation is used, training stimuli are shown in random order. If many examples of one class were shown first to the network, it would learn to recognize these; but when they were followed by many examples of a second class, what had been learned about the first set would gradually be forgotten.

The reasoning behind backpropagation is that those units which are most responsible for the output signal being wrong are those whose connection weights should be changed by the largest amount.

It used not to be uncommon for backpropagation networks to take days or even weeks of computer time to train. This is now unusual, but large networks may still require several hours of training to reach a respectable level of performance.

A network in which connection weights have been trapped by a local minimum will still function, but less effectively than one in which the global minimum has been found.

It is also possible for 'network paralysis' to set in. As a network is trained, it is common for connection weights to show a gradual increase in size. The size of the adjustment to the weights during backpropagation is proportional to the derivative of the sigmoid function, but if the weighted input signals become very large, this derivative is almost zero (far to the right in Fig. 2.20). The network then finds it hard to determine how connection weights should be adjusted to reduce the error function, and may settle into a state of morbidity in which further learning is virtually impossible.

Despite these limitations, backpropagation is a powerful way to train neural networks. It can generate trained networks superior to conventional programs not only in performance but also in several other respects.

2.7 Advantages of artificial neural networks

Artificial neural networks are well suited to operation in unpredictable and poorly-understood environments. The features which make the human brain particularly potent in learning and reasoning tasks are precisely those which make it a powerful scientific tool, and many of these features can be reproduced in an artificial neural network. Although artificial networks at present cannot approach the wide-ranging abilities of the human brain, they display several advantages over conventional problem-solving algorithms.

Diversity of applications

An artificial neural network need not be designed for a particular application; through suitable training it can undertake a wide variety of tasks.

References to several dozen applications of artificial neural networks are given in the books listed at the end of this chapter.

Training determines the role of a neural network. No network will be expert at both analysing seismic signals and judging literary style, but commercially-available neural network software is often very general in nature, and might equally well be applied to spectral interpretation or to assessing the suitability of a potential home-owner for a mortgage. The range of applications of neural networks within science gives a good indication of how this characteristic is reflected in practice; some of these applications are mentioned in the final section of this chapter.

Fault tolerance

Neural networks are fault-tolerant.

Every day many neurons in the brain die (a few pints of beer or glasses of wine increase the rate of this process), yet the power of the brain appears unimpaired by this loss of some of its processing units.

If the ability to understand quantum mechanics or speak Spanish depended crucially on a single neuron, the skill would be irretrievably lost if that neuron died. Fortunately, no single neuron is of pivotal importance; there is no centralized processing point, and the loss of a small proportion of their number does not bring processing to a halt. The gradual loss of neurons is regrettable, but there is sufficient flexibility in the construction of the brain that their role can be assumed by other neurons without any noticeable diminution in performance. The brain is therefore **fault-tolerant**.

As processing units are lost, performance diminishes gradually, rather than catastrophically. This behaviour is known as **graceful degradation**, and is a characteristic also of artificial networks. Of course, we are unlikely to 'lose' units from an artificial network, but graceful degradation applies equally well to loss of quality of the input signals. If these signals become noisy, or some are lost completely, an artificial network will continue to work, though its answers will contain a greater degree of uncertainty. This is a useful feature when scientific data such as blurred photographs or noisy spectra are being assessed.

Tolerance of damage in the brain is further enhanced by the fact that it appears to store the same memory redundantly in several locations, which is a clear protection against loss of neurons. This feature too may be reproduced in artificial neural networks.

Fault tolerance is a notable characteristic of any network, artificial or biological, and this was vividly demonstrated by one of the first artificial neural networks, built in 1951. This novel construction contained hundreds of vacuum tubes, numerous unreliable soldered connections, knobs, dials and a gyro pilot from a Second World War bomber. Because of the unreliability of its components and wiring, the system was often not fully operational. Despite this, it was able to function even when some of its 'neurons' worked incorrectly, or not at all. In all neural networks, loss of some processing units causes a degradation in performance; unless the damage is severe, however, such loss does not precipitate complete failure.

Ability to deal with new situations

Once trained, a network can deal with previously unseen data.

We know what a crystal looks like; we are 'trained' to recognize them because we have seen many examples before. If you were given some uranium nitrate, you could quickly assess whether or not it was crystalline, even if this was the first uranium compound you had seen. The brain stores generalized notions of concepts such as 'crystal', or 'rice pudding'; this allows us to recognize examples of them, even if they are different from every other example we have encountered.

This is precisely the way an artificial network operates. The hidden units in the network organize themselves into groups responsible for recognizing certain key features, detecting when a particular combination of input units is on. If enough hidden units exist, the hidden layer may form an **internal representation** of external patterns and, when they recognize the pattern, pass the appropriate response on to the next layer. If this layer is a further hidden layer, it may in turn determine what *combinations* of patterns detected by the first layer are present. This allows the network to form generalized notions of patterns and their combination, which is a skill of importance in areas from science ('What does an abnormal white blood cell look like?') to robot vision ('Is this a hand I see before me?')

The correct interpretation of previously unseen images is a vital ability of intelligent creatures, and one on which the usefulness of a neural network in many applications depends most markedly.

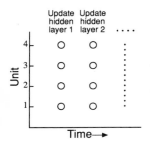

Fig. 2.21 Parallel updating of connection weights in a neural network.

Computers are now being designed in which the neural network is integrated into the circuits of parallel processors themselves. A neural network implementation of this sort is particularly efficient, and represents a step towards a true artificial brain.

In some fascinating work, neural networks have been trained on data which scientists have used in the past in formulating new scientific laws. Analysis of the conclusions of the network shows that, in many cases, they independently discover the same laws underlying the data that human scientists find.

Parallel operation

Artificial neural networks have the capability for parallel operation built in.

Many modern computers contain multiple processors (from two processors to many thousand). In these machines the calculation is divided among different processors, in a type of working known as parallel processing.

The brain operates as a parallel device, and since the capability for parallel working is 'built into' artificial neural networks, adapting them to run on parallel machines is relatively straightforward. The computations that take place at one unit are independent of what is happening at other units in the same layer (Fig. 2.21), so the updating of connection weights at all units in a single layer can be carried out simultaneously if each unit is allocated to a different processor in a multi-processor machine. This can lead to a huge increase in the speed at which the network can be trained and operated.

Rule discovery

Artificial neural networks operate by discovering new relationships among their input data.

Neural networks are particularly adept at discerning regularities in data. A network trained on scientific data will form an internal representation of rules which allow it to correctly interpret the data. These rules may be an empirical interpretation of scientific rules with which scientists are already familiar, but there is no reason why they must be. The network may equally well discover relationships between the data ('scientific laws') which were previously unknown. This is an exciting development, with enormous implications for science.

The interpretation of these new rules presents a serious challenge, though, as you might already have appreciated: the rules will be buried in connection weights between artificial neurons, the meaning of which must be decoded if the new rules are to be understood in scientific terms. The knowledge in any neural network is not very accessible (or more accurately, it is accessible, but obscure), being tied up in these connection weights. The thinking of the network cannot readily be understood by just inspecting the weights.

There is a marked contrast here between a physical 'black box' whose purpose is to perform a fixed scientific task such as controlling conditions in a chemical reactor, and a stable neural network trained to complete the same task. If a black box is opened up, an expert will recognize the electronic components within and be able to understand their purpose, and why the box functions as it does.

If one 'opens up' a trained network, all that is revealed are the connection weights and threshold or bias values, which give no simple guide to what the network is designed to accomplish, or how it does it. Neural networks are in this sense oracular, that is, the output from a trained network is related in convoluted ways to its input. (There was no way to know if statements from the Oracle at Delphi were correct or not. In a similar fashion, one must trust that a network trained to deal with a complex problem has learnt correctly and knows what it is doing. While a network may arrive at the correct answer, it is unable to describe how it got there.)

The task of unravelling connection weights to interpret laws discovered by the network is challenging, but the potential is very great, and soon artificial networks may be helping not just in the analysis of scientific data, but also in the derivation of the laws that explain the data.

Incorporation of fuzzy logic

Artificial networks are, because of their design, able to cope with 'fuzzy data' without modification.

Fuzzy logic is used to analyse data which contain uncertainty or are based on judgement. Strictly, fuzzy data are those in which the transition from one state to another is gradual, rather than sharply-defined. A 'slow car' does not become a 'fast car' at some universally-agreed speed, so the designation of a car as 'fast' or 'slow' is fuzzy.

Suppose a customer of a water company complains that her tap water tastes of chlorine. How much chlorine is present? There is not much to go on, but fuzzy logic can extract low-grade quantitative information from such data. If a customer complains that water is 'unpleasant' or 'tainted', there is probably less dissolved chlorine than if it is 'disgusting', or 'like a swimming pool'. The lowest concentration of chlorine in water that can be detected by most people is around 0.1 to 0.6 mg l^{-1} (the level depends upon whether you have been eating spicy food, whether you are a smoker, and other factors). A swimming pool contains around 3 mg l^{-1} and at concentrations above about 10 mg l^{-1} most people find water undrinkable. These figures place limits on the amount of chlorine present and, if they are combined with the customer's subjective assessment of what the water tastes like, an approximate value for the concentration can be chosen. This combination of numerical with subjective data is the essence of fuzzy logic.

Subjective descriptions, such as 'large', 'sticky', or 'bright' can be fed into an artificial neural network by asking the user to define numerically what 'large' or 'sticky' means to them. Even though the training stimulus might be defined in fuzzy terms, for every stimulus there is a *precise* training target, so woolly data can be fed into a neural network and definite (and, one hopes, reliable) conclusions fed out. The ability to handle fuzzy data is a great asset, and, since neural networks have this ability built-in, no special modification needs to be made to a working network to allow it to handle fuzzy data, such as human speech, handwriting or data containing uncertainty.

The oracular nature of networks can lead to some curious problems. A military neural network was trained to recognize tanks by showing it photographs of a tank in the countryside, sometimes in the open, at other times partly hidden behind trees or other obstructions. The training set included further photographs of countryside without a tank to be seen. In due course, the network learnt to distinguish between the photographs that showed the tank, and those without it. After this apprenticeship, it was shown a new set of photographs, with the tank in different surroundings. Unexpectedly, the network failed to detect even the most obvious of tanks. Eventually it was noticed that every photograph of the tank in the training set had been taken on a day of bright sunshine, while all other photographs had been taken on a cloudy day. The network had learnt to recognize when the sun was out.

Entering fuzzy data into a network is easier than it may sound. For example, 'On a scale of one to ten, how fast is a Reliant Robin?'

2.8 Hopfield nets

Multi-layer networks are not the only type that may be used to assess
scientific data. In a Hopfield network, every node is connected to every other
node (Fig. 2.22); this is a **self-organization model**.

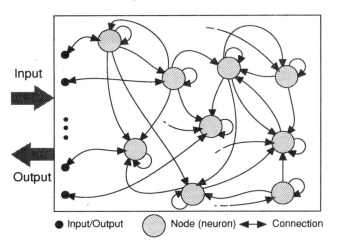

Fig. 2.22 A Hopfield network.

If the connection weights
between nodes in a Hopfield
network are not symmetrical, the
network may never converge to
a stable state and may, instead,
oscillate unpredictably. Such
oscillations have been studied
by workers interested in chaos.

There is no separate input or output layer; instead every node receives
input signals from the environment, and every node has an output to it.
Connection weights between each pair of nodes are symmetrical; that is, they
are equal for messages passed in either direction.

The manner in which a Hopfield network operates is necessarily rather
different from a multi-layer network. Input signals from the environment are
applied to all nodes simultaneously. Random starting connection weights are
used to generate an output signal and this is immediately fed back to all nodes
as a new input. The process is repeated until the network reaches a stable state
and further cycles produce no change. The final outputs from the nodes are
taken as the response of the network.

The trained network contains multiple patterns stored in the coded form of
the connection weights. The stored pattern that is closest to an input pattern
becomes the output pattern. This is then a type of **associative memory**; an
input consisting of just part of a stored pattern can trigger regurgitation of the
complete pattern by the network (just as a long-forgotten tune or aroma can
bring to mind a flood of associated memories.)

Hopfield nets can learn successfully using a step function instead of a
sigmoidal function for thresholding, since backpropagation is clearly
inapplicable in a network of this topology. They are prone, however, to falling
into solutions which are sub-optimal, and special measures may be taken to
prevent this happening. These measures, such as simulated annealing (which
uses a 'temperature' term and a Boltzmann factor which will be familiar to
chemists), are discussed in the books on neural networks listed in the
bibliography.

2.9 Applications of artificial neural networks

Neural network programs are available for computers from desktop PCs to supercomputers. In principle, these programs may be applied to any problem in which there is a mapping between an input vector and an output vector, but, as one might expect, they are more suited to some types of problems than others. They are at their most valuable when applied to problems in which:

- Data are incomplete or unreliable, so that a deterministic program would find it difficult to reach a solution.
- Numerous training examples are available.
- The rules which define how a given input triggers a particular output are incomplete or unknown.
- An explanation of how any decision is reached is not required.

The range of problems to which neural networks have been applied is considerable, and there is space here only to mention a few. Networks are being used in projects to model non-linear chemical systems, such as continuous-flow stirred tank reactors, to predict the secondary structure of proteins and to identify coding sequences in DNA. Since they are good at recognizing patterns buried in noise, they have been used to filter data from experiments in which pulsed lasers trigger chemical reactions, and in the medical field, to analyse patterns of nerve pulses, and to monitor the heartbeats of patients in hospitals, watching for irregularities or unusual patterns in electrocardiograms.

If sufficient examples are available, large neural networks can learn to recognize patterns of arbitrary complexity. They can then be used as pattern detectors in applications from the analysis of magnetic resonance brain images or the interpretation of electron microscopy scans, to the visual assessment of parts emerging from a manufacturing process, or seismic traces. Neural networks are also being used by banks in trials of cheque signature recognition, and by the Post Office in interpreting handwritten postal codes.

The analysis of near infrared (NIR) spectra is difficult for conventional methods, such as partial least squares, because there is a non-linear relation between transmittance and concentration. Neural networks are untroubled by this and are particularly valuable in NIR spectroscopy, where they are now widely used.

Neural networks have been used to predict the 'magic islands' of nuclear stability that are believed to exist for superheavy atoms. This is an intriguing application because it appears to require the network to extrapolate from a region of known stability to an unknown region and, as you will recall, artificial networks are poor at extrapolation. To avoid this difficulty, the problem was re-cast so that the network was trained to move *between* regions of stability. An extrapolation problem was thus turned into an interpolation problem.

Since in production mode, signals pass in one direction only through the network, even a large trained network can make rapid decisions. Consequently, networks can also be used for tasks in which speed is important. Examples include the use of real-time Raman spectroscopy to monitor chemical process lines, on-line analysis of high-energy events in particle accelerators and the construction of three-dimensional displays of electron microscope images.

Because of their generic nature, neural networks are easy to use, and, once trained, require little operator skill. Their ability to deal with data related by ill-defined rules gives them particular power in just those areas which are hardest for conventional programs. As familiarity with artificial networks grows, they will become increasingly common controlling experiments, analysing experimental data derived from them and, ultimately, deriving some of the scientific laws that describe the data.

Further reading

Beale, R. and Jackson, T. (1990). *Neural computing: an introduction*. Adam Hilger, Bristol.

Freeman, J.A. and Skapura, D.M. (1991). *Neural networks: algorithms, applications and programming techniques*. Addison-Wesley, Reading, Mass.

Lisboa, P.G.J. (ed.) (1992). *Neural networks current applications*. Chapman and Hall, London.

Nelson, M.M. and Illingworth, W.T. (1991). *A practical guide to neural networks*. Addison-Wesley, Reading, Mass.

Rumelhart, D.E. and McClelland, J.L. (1986). *Parallel distributed processing: explorations in the microstructure of cognition*. MIT Press, Cambridge, Mass.

Wasserman, P.D. (1989). *Neural computing theory and practice*. Van Nostrand Reinhold, New York.

3 Expert systems

3.1 Introduction

An expert is a valuable commodity, especially in a field such as science which requires a high level of technical understanding. However, experts are not always available in the laboratory, and their expertise sometimes lies in areas beyond what is needed. Because of this, scientists are being advised – and sometimes replaced – by inanimate helpers, known as expert systems.

An expert system is a computerized clone of a human expert.

Expert systems contain a fund of knowledge relating to some specific, well-defined area. By using rules to combine this knowledge with information gathered from the user, they draw conclusions, provide advice, and help the user to choose between alternatives. Through the medium of an expert system, a computer can reproduce many of the logical deductions of a human expert and offer scientific advice of a quality rivalling that which a human might provide. If autonomous, expert systems may control a spectrometer, a missile guidance system, or a nuclear power plant, without outside assistance.

When you work with an effective expert system, it is much like having someone beside you with whom you can discuss the problem that you are trying to solve, and how you propose to solve it. Through a two-way conversation, the expert system guides you in the development of an organic synthesis, control of an industrial plant, or other complex task. In most respects the expert system behaves like a human and asks the kinds of questions a human might ask. However, this expert is confined to the inside of a computer.

3.2 The need for expert systems

In this chapter we shall outline how expert systems operate, and how they can be applied to science. We shall discuss how computers can reason and draw conclusions, mimicking what is commonly called 'thinking'.

The potential for expert systems is particularly marked in science. Scientific experts are not a rare breed of course, but the amount of scientific knowledge is growing so rapidly that it is becoming difficult to know everything of importance in even a tiny area of specialization. Most scientists who want expert advice on an occasional basis do not have the time to maintain expertise in all the relevant areas; a computerized expert may then be the only feasible way to get help when it is needed.

been at the centre
velopment of expert
ns for years; one of the
iest and most influential of
such systems was written
specifically to help chemists.
This was the program
DENDRAL (developed from
1965 onwards at Stanford
University), whose role was to
propose possible structures for
organic molecules through
interpretation of their mass
spectra.

As a result, expert systems, which are already the most widespread application of AI in science, are certain to become much more common. Since the appearance of DENDRAL in the mid-1960s, expert systems have multiplied to encompass a broad range of applications, and they have potential in virtually every field in which expert advice is now provided by humans. They are already used in such areas as genetic engineering, analytical chemistry, chemical safety and experiment design, and the range of applications continues to grow rapidly.

3.3 The components of an expert system

We consider first the components and characteristics of an expert system. What makes a computer program so good it deserves to be called 'expert'?

Early expert systems contained in condensed form both the questions that a user might want to ask about a particular topic, and appropriate answers; there was also a rudimentary means of interpreting the users' requests. These programs were small electronic repositories of facts, and had only a glimmer of intelligence within. No consideration went into the computer's response, there was no 'mulling over' by it of what the user wanted or how it might meet the user's need. The computer had a fixed response (or no response) and that delimited what advice it could offer.

A human expert is very different: he or she can respond *intelligently* to queries. Any computer program with pretensions to being called 'expert' must be able to do the same. It must be knowledgeable and lucid about its own field, but flexible enough to respond constructively to questions at the fringes of its expertise, or questions that the designer of the system had not anticipated. This clearly requires more than just the ability to sift through a prepared list of responses.

Put simply, an expert system must be *thoughtful*.

The key to an expert system therefore is its ability to reason and to make judgements; this is the essence of being able to act thoughtfully. Many human tasks involve judgement: in applications such as real-time process control or experiment design this is crucial. These are typical of areas in which an expert system is valuable, and decisions about them must be made based on the interpretation of a pool of specialized knowledge; effective decision-making requires access to all the relevant information.

However, as we have suggested, to be an expert one needs more than just a list of facts, and a modern expert system is composed of three essential components (Fig. 3.1).

The **knowledge base** contains information about which the system reasons. This permanent memory contains as much information as possible relevant to the field in which the system is 'expert'; during operation it is supplemented with transient data supplied by the user and perhaps gathered from the environment.

The knowledge base might, for example, contain details of the thermodynamic properties of small molecules, or a description of the identity and positions of the components of a spectrometer, but it is unlikely to contain information about several completely unrelated areas.

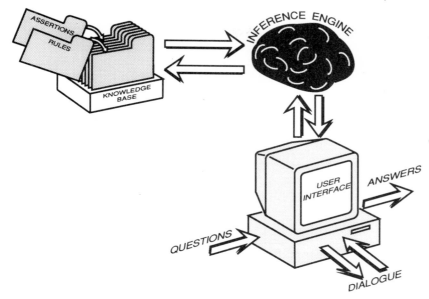

Fig. 3.1 Components of an expert system.

Most humans have broad general knowledge, but are experts only in a narrow, well-defined field. In a similar way, expert systems restrict their attention to a specific domain, such as insurance risks or chemical pollution analysis, rather than the broader domain of life, the universe, and everything.

Because of its reliance on stored facts, an expert system is an example of a **knowledge-based system**. These systems use 'rules of thumb' in their operation, rather than numerical algorithms, and work with information which has been encoded so that it can be stored conveniently in the computer. A knowledge-based system is at its most valuable when dealing with large and complex databases, or controlling equipment with many components whose operation is governed by numerous inter-related rules. Simple problems do not require complex solutions: an expert system would be wasted running a coffee machine or an electric stapler.

However, in complex problems, there is naturally a great deal to know, and though expert systems are often at their most productive when working on these problems, they cannot know everything. Accordingly, the knowledge base will reflect a focused approach and contain essential information only.

The **inference engine** is the part of the system that reasons or infers. Its responsibility is to draw conclusions, using data, rules, and relationships from its knowledge base, and whatever it has learned through conversation with the user, and to justify those conclusions.

Decision-making by the inference engine is a fundamental part of the operation of an expert system; indeed it is *the* fundamental part, since unless it can make decisions, the expert system will be valueless. An expert system replaces or advises a human expert by *reproducing the results of his or her reasoning*, so if the artificial expert is given the same information as the

The terms 'expert system' and 'knowledge-based system' are now used almost interchangeably. Early programs were entirely dependent on input from a human expert, (and most still are), so they became known as 'expert systems'. Some modern systems gather much of their information autonomously, without the need for a human expert, and the term 'knowledge-based system' is then more appropriate.

It is not essential that the inference engine use the same method of reasoning as a human expert, just that the final conclusions be the same.

human, it should reach the same conclusions (or, if it does not, it should be able to show in what respect the human is wrong).

The **user interface** links the system to the outside world. Through this, the user asks questions and provides information, receiving in return advice, comments, and conclusions. Some expert systems control part of their world, such as a chemical synthesis line in an industrial plant. The expert system then has interfaces (message-passing links) to motors, detectors, bar-code readers or other equipment, whose purpose is to transfer information about the environment into the system, and provide a mechanism to implement any actions that the system thinks are necessary.

These three parts of the expert system, the knowledge base, the inference engine, and the user interface, are both conceptually quite distinct and logically separate. They function simultaneously but independently, swapping bundles of information as required. Because of this independence, an expert system can be built incrementally; one part can be modified or augmented without affecting the others. Working expert systems are large and complex; division into discrete units simplifies testing when the system is being built, and permits periodic updating.

We now consider in more detail each of the three components in an expert system.

3.4 Information and rules: the knowledge base

Expert systems interact with the user or the environment to find out what must be done. Some data that the system needs are gathered through this conversation, but most are held in the knowledge base, which is the prime source of information that the expert system needs to reason. It contains **facts**, **rules**, **relationships**, and **statements of probability**.

In the simplest of knowledge bases, information is stored as an unordered list (Table 3.1), through which the expert system searches each time a query is presented. This is a crude method for holding information, since the knowledge base may be large, and contain possibly thousands of entries inter-related in arbitrary ways. If the list is organized without a clear structure, it will be hard to locate relevant data, so it is desirable that the information is arranged logically according to some suitable prescription.

Often, the data bear a relationship to each other that allows them to be arranged on a tree diagram, like that in Fig. 3.2. The top of the tree is known as the **root**; the connections are **branches**, and all extremities other than the root are **leaves**. Each leaf represents one or more items of information (or possibly instructions on how to obtain information).

This information will not be stored literally as words and phrases, but in a coded form to increase storage efficiency and to simplify the interpretation of rules by the inference engine. Symbols are used to represent concepts and the relationships between them, and it is the task of a **representation language** to organize this mix of rules and data in a coherent fashion.

Table 3.1 Part of an unstructured knowledge base.

1. Pooh is a bear.
2. Wol cannot spell.
3. Eeyore's tail is nailed on.
4. A burst balloon will fit inside a honey pot.
5. Pooh has very little brain.
6. . . . . .

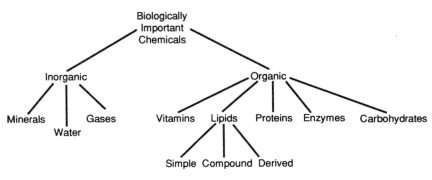

A typical entry at the 'carbohydrates' leaf in Fig. 3.2 might be FORMULA= CARBOHYDRATE[$C_x(H_2O)_y$]. This would be held in coded form for efficiency of storage and use by the inference engine.

Fig. 3.2 A data tree.

The hierarchical structure of the tree shown in Fig. 3.2 reminds one of a family tree, and an important feature of this type of structure is that it implies **inheritance**; in other words, entries in the tree inherit the attributes of entries above them to which they are linked by branches. This simplifies the reasoning process which the inference engine employs, since, by working down from the root, it can quickly narrow its search for the data that it needs by following appropriate branches.

A tree is a simple and effective rubric for organizing information, but not all types of data can be classified in this fashion. If the data are inter-related in a complex manner, a hierarchical format is inappropriate, since items connected by logical links may share different relationships. Data may then be arranged in a more disorganized-looking network (Fig. 3.3), in which branches link entries wherever a logical connection exists.

'Compound' and 'derived' are both types of lipid, and lipids in turn are organic molecules; this inherited relationship is clear from the structure of the tree in Fig. 3.2.

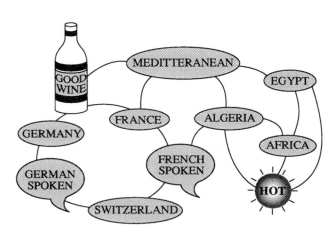

Fig. 3.3 A disorganized network.

Pieces of information within the knowledge base are known as **assertions**. A probability may be associated with the assertion if it is known, and, when present, will be taken into account by the inference engine when the assertion is used. Many assertions are 'true'; for example,

$$POINT\_GROUP = AuCl_4 [D_{4h}, 1.0]$$

states that the point group of gold(IV) chloride is D_{4h} (definitely, since the probability that the assertion holds is 1.0). Other assertions hold with some smaller degree of probability:

$$MCDONALDS = IN[LARGE\_TOWN, 0.85]$$

Although the McDonald's assertion is complete, it relies on other information, since the inference engine must know what is meant by a 'large town'. The knowledge base might contain a suitable definition. If it did not, when the expert system was responding to a request ('Where's the finest food in Paris?'), and it noticed this assertion, the inference engine would ask the user to define a large town, or to give examples, so that the expert system could learn what one was like.

tells us that there is an 85 per cent chance that a large town has a McDonald's restaurant.

The knowledge base can contain assertions about numerical data, formulae, text, lists or tables, and physical descriptions; any information, in fact, that might be relevant to its area of expertise. In turn, this information may be fixed, for example a list of atom electronegativities, or dynamic, such as the current refractive index of material in a reaction vessel.

Assertions are the raw material about which the expert system reasons. They form the heart of the knowledge base, but if this were composed of assertions alone, the expert system would be merely pages within a computerized encyclopaedia, and this hardly qualifies as an intelligent system. Therefore, in addition, the knowledge base contains prescriptions or rules which instruct the inference engine on how to combine assertions. A solid mathematical base underpins these prescriptions, but we shall restrict our discussion to non-mathematical aspects of their use.

3.5 Reasoning: the inference engine

The inference engine does the 'thinking' for an expert system. It scans the knowledge base for assertions and rules relevant to the user's question; any pertinent rules are used in an attempt to draw useful conclusions. The assertions that it discovers may be 'facts', which can be combined to discover what is 'true'. If the assertions have probabilities of less than 1, conclusions can only be drawn with some degree of confidence.

A 'fact' has a probability of 1.

The recipes for combining assertions take a number of forms, including **predicate logic**, **structured objects**, and **production rules**. Production rules are the most widely used recipe in chemistry and are the most readily understood; they take the form

A typical production rule might be:
If the matrix is mine tailings, **and if** a test for gold is positive, **then** cyanide will be present **and** further tests should be carried out in a fume cupboard.

If assertion$_1$ holds **and** assertion$_2$ holds **and** .....
...... **then** draw conclusion$_1$ **and** draw conclusion$_2$ **and** ....

Production rules can generally be written as a statement in English, which makes them straightforward to understand (Table 3.1)

Table 3.1 A production rule from an expert system in 'pseudo-code'.

```
begin
   rule
          'If the levels of both ethane and propane are high,
          then take action to reduce ingress of pump oil.'
   endrule
          if    (ethane  > acceptable_ethane_level)
          and (propane > acceptable_propane_level)
          then
          advise
                 Nature of problem: pump oil ingress.
                 Danger to system: possible sample contamination.
                 Required action:   renew pump seals.
          endadvise
          put ('yes' into located_problem)
          reset (more)
          endif
endprocedure
```

The knowledge base will contain many production rules. Some of these will be very specific, ('**If** the indicator must change colour in the pH range 1.2 – 2.8 **then** use methyl blue'), while others will be more general ('**If** the molecules are associated in neither the liquid nor the gas phase, **then** Trouton's rule should apply').

The rules are generally entered by a human expert as the system is being built, but this is not always possible. Sometimes it may be hard to mould the relevant knowledge into concrete rules, or there may be large amounts of data available, but the rules that express in general terms the relationship between different items of data may be unknown. In these circumstances, the expert system may itself derive production rules and the probabilities within them by examining examples that illustrate application of the rule, provided by the human expert. This method of teaching the expert system is known as an **induction system** or **automated knowledge elicitation**.

Most knowledge bases contain at least several dozen rules; sometimes many hundreds will be present.

When the rules expressing the relationships that exist among the data are poorly understood, a neural network may provide an alternative to an induction system.

Searching strategies

An expert system can only be really effective when the knowledge base contains all the information relevant to a problem. And yet, when a particular problem is tackled, most information in the knowledge base is irrelevant. The efficiency of the inference engine is therefore strongly influenced by how quickly it can locate data which might bear upon a problem. It needs

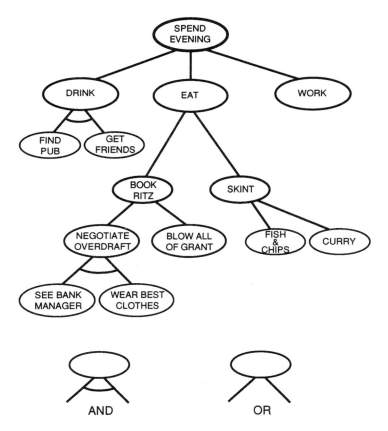

Fig. 3.4 An AND/OR tree.

some kind of strategy for doing this, otherwise it may stumble through the knowledge base without direction, and be unworkably slow.

A widely-used method for drawing the attention of the inference engine to relevant information is to use a tree of goals to guide its reasoning, like that shown in Fig. 3.4.

This is an **AND/OR tree** or **decision tree**. The root defines a goal which will be met if a certain combination of sub-goals is met. Sub-goals lie below the root, and are joined to it and to other sub-goals, by branches.

Some leaves of the tree are not sub-goals, but are data that cannot be deduced and may or may not currently be known. The inference engine may interrogate the knowledge base to find out whether the data at these leaves are available or can be calculated, or the data might be supplied through a dialogue with the user. Alternatively, the leaves might represent instructions whose execution could yield data. For example, examination of a leaf might trigger a request for an experimental operation to be performed, such as the measurement of the pH of a solution, and the reporting back of that pH to the inference engine.

Branches from a goal to sub-goals below it may be dependent or independent. Some goals are met only if multiple sub-goals are satisfied simultaneously (AND conditions); branches to these sub-goals are connected by an arc. Thus, Fig. 3.4 reveals that to negotiate an overdraft, it is necessary to arrange a meeting with the bank manager *and* to wear one's best clothes.

On the other hand, a goal might be met if just one of its sub-goals holds. The evening can be spent drinking *or* eating *or* working. The drinking, eating, and working sub-goals are independent of each other and represent OR conditions. (This of course does not prevent more than one sub-goal being satisfied – we might both eat and work.)

The purpose of the decision tree is to represent production rules in a logical way. There is a simple relationship between a decision tree and a set of production rules. Suppose the rules shown in Table 3.2 apply. This set of rules can be represented by the decision tree shown in Fig. 3.5.

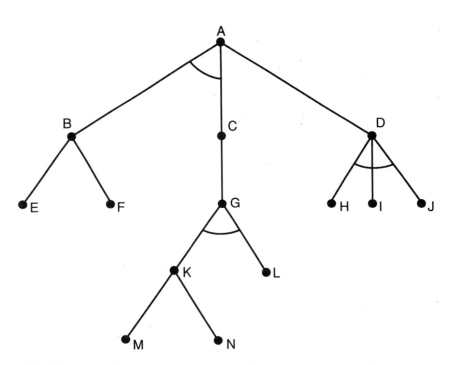

Fig. 3.5 A decision tree for the set of production rules shown in Table 3.2.

Table 3.2 Production rules.

- If B and C hold, or D holds, then conclude A.
- If E or F hold, then conclude B.
- If G holds, then conclude C.
- If H and I and J all hold, then conclude D.
- If K and L hold, then conclude G.
- If M or N hold, then conclude K.

Using a representation of this tree within the knowledge base, the inference engine tries to prove its goal. The tree shows that, if A is to be proven, either B and C must simultaneously hold, or D must hold. As B, C and D are all sub-goals, not leaves which can be inspected immediately, the tree must be searched more deeply to establish whether A is true.

Although we have hardly started on the task of proving the goal, already a question of strategy arises: should all branches at one level be inspected before moving deeper, or instead should a single branch of the tree, such as that running through node C, be followed down to its leaves before any other branch is considered? These two alternative methods of analysis are known as **breadth-first searching** and **depth-first searching**.

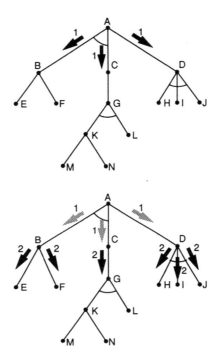

Fig. 3.6 A breadth-first search.

Breadth-first searching

In a breadth-first search (Fig. 3.6) all branches to sub-goals or leaves at a given level are assessed together. There may be leaves at this level which can be inspected to see if the goal of the tree is proven. If there are no leaves, or the goal cannot be proven, the inference engine turns its attention to the group of branches at the next level down, and the process is repeated, level by level, until either the goal is proven, or all branches have been investigated.

The advantage of this type of search is that the user can keep track of the line of reasoning that the expert system is following, as it narrows down the range of possibilities, and the user can intervene if necessary to guide the search along the most productive path. For example, in the chemical analysis of a sample of a completely unknown material, a continuing conversation between the user and an expert system will permit the chemist to combine her own and the expert system's knowledge to greatest effect.

Breadth-first searching invariably finds the shortest path from the root to a solution, but it does not always find a solution more rapidly than other searching methods, since the shortest path can only be found by exhaustive checking of each layer in turn, and if the decision tree is large this can be time-consuming.

Depth-first searching

In a depth-first search (Fig. 3.7), a single branch from the root is pursued to its leaves. If the goal is not proved when the data that these leaves supply are assessed, the inference engine moves down through the neighbouring branch to its leaves, and the process is repeated until either the goal is proved, or the branching structure is exhausted and the inference engine concludes that the goal cannot be met. Buried within the knowledge base may be hints that certain branches will be more promising than others, and if the inference engine can decide intelligently which branch to pick next, this is normally a faster method of searching than breadth-first.

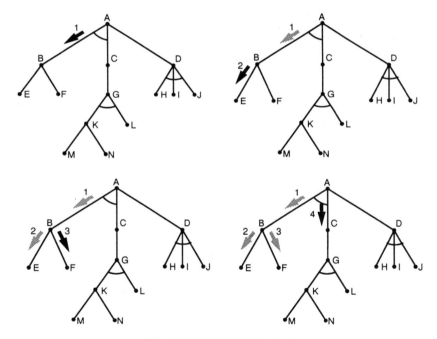

Fig. 3.7 A depth-first search.

Both depth-first and breadth-first searches report back to the user as soon as a successful conclusion is reached. If the user wishes to see other solutions, the system must be told to continue the search further. From Figs 3.6 and 3.7 you can appreciate that, when several solutions exist, they will usually be discovered in a different order using breadth-first and depth-first search. For example, a search through a library of infrared absorption spectra for a match to an experimental spectrum never uncovers a perfect fit; slight differences in band position or absorption intensity always exist. Consequently, spectral matching programs offer several candidate formulae as possible matches and, if an expert system plays a role in the selection of these, matches may be discovered in different orders, depending on whether depth-first or breadth-first searching is chosen.

Backward- and forward-chaining

Breadth-first and depth-first searching are types of **top-down inference** or **backward-chaining**. The inference engine starts from a hypothesis or goal, and reasons down through successive layers of sub-goals until it reaches leaves which it can examine to see whether they hold. The goal itself can then be shown to hold or to be contradicted.

This is an efficient way to reason since, as soon as a particular sub-goal has been contradicted, any further sub-goals branching off it can be ignored, so wasteful searching is avoided. However, backward-chaining suffers from a limitation which restricts the types of application in which it can be used: if we do not know what question to ask of the expert system, we can not even start.

You might think that this should not worry us. After all, we do not usually begin to search for something until we know what it is that we are looking for. Nevertheless, there are scientific searching problems in which the goal can be stated in only the most general terms.

Suppose a knowledge base contained all available knowledge about the chemical and physical properties of alloys; there might be details of composition, density, melting point, entropy of fusion, and so on. The role of the expert system which contains this knowledge is to scan the information, looking for relationships that are not contained as assertions or rules in the knowledge base. Since the knowledge base contains all that we know, these new relationships are therefore empirical rules of whose existence we are unaware. Perhaps the melting point of a certain type of alloy correlates with ductility, or the phase diagram of an alloy depends in some as yet unrecognized way on the van der Waals radii of the component atoms. Waiting in the data there are scientific rules to be discovered, if the expert system can only find them for us.

Because we do not know what relationships may emerge, beyond those that are already encoded as assertions, the goal of the inference engine cannot be stated much more precisely than 'look for new relationships'. It is hard to see how this could be broken down into sub-goals to form a decision-tree, unless there were a separate sub-goal for every possible relationship, and in a large database this would lead to a massive decision tree. Backward-chaining cannot be used.

The solution is to use **forward-chaining**. In forward-chaining the inference engine extracts successively more information from the user and knowledge base and combines this using its rules to generate new conclusions or assertions; these are then examined for value by the expert system or the user. Once again some mechanism (perhaps involving the user directly) may

The sort of analysis described here is taking place in several universities. Huge stores of information are searched by expert systems looking for previously unknown relationships among the data. Currently such research is centred on statistical and sociological databases, and is yielding some interesting, though often obscure, relationships. There is, however, every prospect that, when applied to scientific data, such projects will discover valuable new relationships.

be needed to focus the attention of the inference engine on the group of assertions and rules of greatest relevance.

Both forward-chaining and backward-chaining are deterministic in the way that they operate. A third method of working exists, which leads the inference engine through the knowledge base on a path that may be somewhat more convoluted than these methods. This procedure is smarter, but more difficult to implement. At each stage in **rule-valued searching**, the inference engine tries to gather data that will most effectively reduce the number of options available at that point. The system attempts to efficiently 'home in on' the correct answer by always looking for the information that will be most effective in reducing the uncertainty in the problem.

Commercial organizations are using **data mining** techniques, in which genetic algorithms or neural networks search through the massive amounts of data generated by, for example, computerized supermarket tellers, and look for hidden trends or correlations. Similar methods are applied to stock market analysis, assessing share price movements. The performance of stock selection software using these techniques is claimed to already exceed that of a human trader by a factor of two.

Table 3.3 A comparison of forward- and backward-chaining.

Backward-chaining	Forward-chaining
Hypothesis: **Use** a zirconium releasing agent,	*Rule:* **If** a sample for AA analysis contains both calcium and phosphate,
Rule: **If** the sample is to be analysed using AA, **and** contains both calcium and phosphate.	*Conclusion:* **then** use a zirconium releasing agent to eliminate phosphate interference.

This can be a potent way of working. Consider the selection of parameters for an HPLC separation. There might be a choice between reversed-phase or normal-phase column packings, gradient or isocratic elution, detection by UV, conductivity or fluorescence, any one of twenty pure solvents or various mixtures of them, and so on (Fig. 3.8). This gives rise to a vast number of possible combinations of operating conditions.

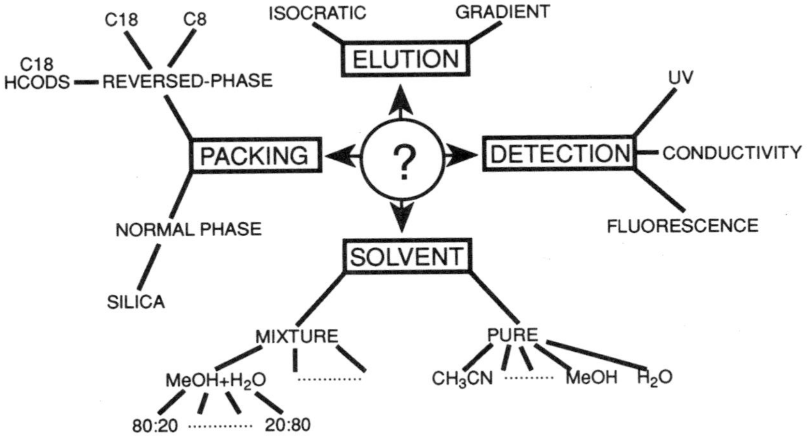

Fig. 3.8 A map of possible operating conditions for an HPLC analysis.

If the expert system first chooses a method of detection, it reduces the number of combinations by a factor of 3 (since one–third of all combinations rely upon UV detection). If, instead, it starts by selecting an appropriate pure or mixed solvent out of perhaps 90 that might be available, the number of combinations is cut by that factor of 90. The size of the problem is thus rapidly reduced if, at each stage in the analysis, the system can make an intelligent choice about what information it should look for next.

Rule-valued searching has clear attractions, since the solutions are being pursued in a thoughtful and direct fashion. However, there are significant overheads. This sort of inference is relatively complicated to implement, and requires constant reassessment of the production rules to determine what should be done next. Gains established by following very direct routes to a solution may be wiped out by the additional time spent inspecting the knowledge base to identify just what those routes are.

As a rule-valued searcher attempts to be efficient, it may introduce considerable variation into the path that the inference engine follows through the knowledge base for only slight differences in input data. The expert system may also pursue a line of thought for a while, only to apparently lose interest in it, before returning unexpectedly to the same line of questioning later. It may feel that it can solve the problem more directly by doing this, though, as we have suggested, it may not always succeed because of the overheads inherent in the method.

> Rule-valued searching is closer to the strategy that a human expert would use to tackle a problem than forward- or backward-chaining.

3.6 Communication: the human interface

The interface with the user is a significant part of every expert system. You might imagine that the user simply poses a question and then sits back to wait for the answer, but this under-estimates the importance of the interface.

An expert system's task is not necessarily complete when it answers a user's question; it must also be able to explain its thinking, that is, show how it reached its conclusions, and justify its arguments.

This may seem of less fundamental importance than the requirement that the system deliver reliable advice, but its significance may become clearer if we recall the role that an expert system plays. It generally advises a user whose own expertise is inferior to that of the system. The expert system may have, and be able to manipulate, knowledge which lies well beyond the competence of the user. As an immediate consequence, *the user cannot usually assess without further investigation whether advice provided by the expert system is reliable; the advice simply has to be taken on trust.*

This can be a bit awkward. No matter how competent the system, sometimes the user will feel sure that the expert system has got things wrong. Suppose an organic synthesis expert system proposed to synthesize phenol starting from carbon monoxide and water; we would surely view this suggestion with a good deal of (justifiable) scepticism. If some convincing

> It is evident that the interface must be at an appropriate level: a human expert, who has asked a computerized expert to confirm his own opinion, will expect a higher level of technical detail than a novice user will need. Equally, while a regurgitation of rules from the knowledge base might be sufficient explanation for a human expert, this would not be a suitable response to queries from a non-expert.

explanation of how the conclusion had been reached, and why it was sensible, was forthcoming, we might at least consider the recommendation, but in the absence of adequate justification, we would seriously doubt whether the expert system understood any organic chemistry. If the interface is sophisticated, the user may be able to engage in a dialogue of investigation; this dialogue can explain how decisions are reached, and will enhance the user's confidence in the system. If the interface is unable to support such a dialogue, even well-founded recommendations of the system may be dismissed by the user.

Working expert systems do not often make such eccentric suggestions, but their conclusions may not always be what the user expected. It is important therefore that the workings of an expert system should be 'transparent': that is, it should be straightforward for the user to determine what the expert system is doing and why it is doing it. Although transparency is especially important when conclusions are unexpected, even when the expert system is getting the right answer, the user needs to be reassured that it is getting the right answer for the right reasons.

A well thought-out interface also reduces the chance that the user will introduce faulty or irrelevant data, whose presence might prevent the system reaching the correct conclusion. Here a question is posed by an expert system helping (?) a user to perform an analysis of metals in an aqueous sample:

Expert System: 'I think precipitating solids from this solution will be a
                     problem; do you agree? (Yes/No)'
User: ' ??? '

What does the expert system mean by this? If the user responds 'yes', will the system interpret this as 'it will be hard to get this sample to precipitate' or as 'this sample may form unwanted precipitates'? Presumably the expert system knows what it is doing, but it is unlikely that the user does. This question is drawn from a poorly-designed interface, and the consequence of such ambiguous requests may be that the expert system fails to answer the question, or (more seriously) offers faulty advice which a novice user might accept without question.

3.7 Imprecise data: heuristics

Scientific data are rarely perfect. An expert system may cope effectively with error-free data, but if it is unreliable when the data are incomplete, incorrect or misleading, it will be worthless in an experimental science such as chemistry. Most scientific expert systems will encounter, and must be able to handle, data containing experimental, human, and other error.

This presents a serious difficulty to the inference engine. Not only need it deal with assertions which are often statements of probability, rather than fact, but also with input data that contain errors. Indeed, the situation is even

Uncertainty is widespread in chemical data. For example, a mass spectrum fragmentation pattern might be complicated by the presence of impurities, or peak intensities might be compromised by instrumental errors. An IR spectrum of an unknown might cover only part of the range of wavelengths that an expert system would like to analyse.

worse than this, since the probability in an assertion may itself be only a matter of guesswork. How should the inference engine handle such uncertainty?

Humans use short–cuts, hunches, and rules of thumb in tackling a problem and it is this sort of specialized (and often intangible) knowledge that makes an expert so valuable. Knowledge derived from experience ('This seismic pattern suggests the presence of oil-bearing rocks below') is of little use in conventional programs, but is potentially crucial to an expert system.

If this woolly information can be incorporated into the knowledge base, it will greatly expand the capability of the system, but turning vague ideas and hunches into concrete rules is rarely straightforward. Such knowledge is not numerical (what is a 'fluffy precipitate' in quantitative terms?) and human experts often have an inflated idea of the reliability of their hunches. No matter how valuable such information might be, the expert system cannot use it unless it can be coded in a form suited to inclusion in the knowledge base. The normal way that this is done is through **heuristics**.

A heuristic is a rule of thumb which gives guidance about how to solve a particular problem.

The key word in this definition is 'guidance'. 'Most drugs are molecules of low molecular weight' is a heuristic. It does not define how to synthesize a useful drug, or what structure it must have, but it narrows the field. A heuristic enables a choice to be made among alternatives when there may be no unambiguous theoretical justification for any particular choice. From this heuristic, we can conclude, given a choice between a polymer of molecular weight 50 000 Da and an organic molecule of molecular weight 300 Da, that the latter is more likely to be a useful drug. The heuristic does not give us enough information to be *sure* that such a conclusion is right, though.

Heuristics are held in the knowledge base as assertions or as production rules. These hold with some probability, and can be expressed using notation of the kind introduced earlier:

DRUG=[MW < 500,0.8]
BALD_MAN=[SEXY,0.01]

If the product is black **and** the product is gooey **and** the product was synthesized by a student **then** the student will not make a good organic chemist (with a probability of 0.95).

Generally the knowledge base does not distinguish between assertions whose probabilities are known precisely (because they can be found statistically), those whose probabilities are a matter of guesswork, and those whose probabilities have been derived through the deductions of the inference engine. For example, for a single molecule, pulled at random out of a racemic mixture, we could write

MOLECULE=[d-isomer,0.5]

with confidence, since a racemic mixture contains exactly 50 per cent of each of the *d* and *l* isomers. This assertion has the same structure as one in which the probability is a matter of guesswork:

ME=[1st_class_degree,0.4]

This is an informed guess, and will remain so until we know the exam results, when it will become an assertion of known probability equal to zero or one.

How production rules combine heuristics

It is a simple matter to use production rules to combine assertions for which all probabilities are unity. For example:

If the compound contains only hydrogen and carbon **and if** there are no rings or multiple bonds, **then** the compound is an alkane.

The inference engine needs merely to determine from the knowledge base or the user whether the conditions hold, and can reach an unambiguous conclusion. Matters are more complicated, though, if the conditions to be combined are heuristics. When production rules incorporate probabilities, they are written in the form:

If condition$_1$ holds *(with probability$_1$)* **and** condition$_2$ holds *(with probability$_2$)* **and ....then** draw conclusion$_1$ *(with certainty$_1$)* **and** draw conclusion$_2$ *(with certainty$_2$)* **and .....**

The confidence with which we can draw each conclusion depends on

- the probability with which each condition holds, and
- the certainty with which the conclusion itself follows, if every condition holds.

Clearly the likelihood that all conditions are simultaneously true cannot be greater than the probability that the most improbable one is true. It is common (but not universal) practice, if there is more than one condition, to set the overall probability that *all* conditions hold equal to the probability that the *least likely* condition in the rule holds.

This is the 'strength of a chain is that of its weakest link' argument.

To see how this works in practice, suppose the knowledge base contains the following information:

DIET=[POOR,0.6]
SYMPTOM=[POOR_NIGHT_VISION,0.8]

with the meaning:

- there is evidence that suggests, with a 60 per cent probability, that the patient has a poor diet;
- there is strong evidence (80 per cent) that the patient has poor low-light vision.

We now apply the production rule:

> **If** the diet is poor **and** the patient has poor low-light vision **then** the patient is suffering from night-blindness (with certainty 0.9) **and** vitamin A should be administered (with certainty 1.0)

The conjunction of the two conditions (the probability that they both hold) is taken to be the smaller of the two probabilities, that is 0.6. The probability that the patient has night-blindness is then the product of the probability that both conditions in the production rule hold (0.6) and the probability that, if those conditions hold, night blindness is indicated (0.9), i.e. $0.9 \times 0.6 = 0.54$.

Even a rudimentary knowledge of statistics will tell you that this is not a statistically valid way to proceed. It cannot be correct to ignore the probability of every condition in a production rule other than that with the lowest probability. If two *independent* events have probabilities of 0.8 and 0.6, the probability that they will both occur is $0.8 \times 0.6 = 0.48$, not 0.6.

Nevertheless, many working expert systems use this weakest link rule. The justification for this apparent disregard of good statistical practice is twofold. First there may be insufficient information available to the expert system to know whether two conditions are statistically related (there is a significant chance that the poor low-light vision is a direct result of the poor diet, so the probabilities of the two conditions are not necessarily independent). Since the way we combine probabilities depends upon whether two events are dependent or independent, there is a potentially unresolvable uncertainty here.

Secondly, the use of statistically more rigorous methods often has little effect on the conclusions of the expert system, but slows its decision-making. If nothing is to be gained by trying to apply statistical methods faithfully, they may (with caution) be put to one side.

It is important to understand that *the power and utility of an expert system is determined primarily by the knowledge that it contains, rather than by the sophistication of the statistical methods that it uses to combine heuristics.* An ability to reason in a statistically pure way can never make up for inadequate information in the knowledge base. The reason for this is that the role of expert systems is often to rank hypotheses. Many questions do not have a single right answer, particularly in science; there may be several answers, or even hundreds. The system must not be defeated by this variability, and will generally tackle the problem by determining which among several alternative conclusions is best, not by attempting to determine the exact probability that its favoured conclusion is 'right'. The certainty with which each conclusion can be drawn is usually of secondary importance, provided that the conclusions are correctly ordered.

This is just the way a human expert works. For example, if only IR and NMR spectra of the product of a new synthesis could be obtained, an organic chemist might be reluctant to suggest an identification for the product from the data available. There might be several structures consistent with the spectra; neither computerized nor human expert could then guarantee positive identification. Various possible structures could be put forward in an order of preference, after the degree to which each explained the spectra had been assessed; further tests might be proposed to help distinguish between the candidate structures. The value of this approach is not determined by whether the exact probability that a particular structure is right can be calculated, but by the probability that, on the basis of the data available, the most promising structure is ranked first.

3.8 Working in real time: using an expert system to direct actions

Expert systems are increasingly used to control instruments or direct industrial processes. When they have this kind of responsibility, their duties go beyond that of just providing advice: they must also implement their own recommendations in the world that they control. Furthermore, as they operate in real time, they must make decisions within fixed and demanding time-scales, and this imposes restrictions on the way that they function.

In a complex industrial installation, it is vital that computers can make intelligent decisions within moments of data, such as temperature and pressure, arriving at an expert system computer. There is no time for decisions to be filtered through a human for approval. If an expert system running a chemical synthesis line makes the wrong decisions, or cannot make the right decisions quickly enough, the result may not be just lost production, but a possible disaster.

Intelligent control of instruments in major industrial plants is growing, because the speed at which information is generated outstrips by a large factor the rate at which humans can understand and process it. All operations of an expert system controlling an industrial plant must therefore be fast, and to guarantee this, it is common for large expert systems to be selective in the data that they gather. The more important data emerging from a chemical plant may be polled continuously at high speed, while other data are monitored in a 'background' mode, the expert system collecting them infrequently, and assessing them when time is available. This mirrors closely what a human operator would choose to do, paying close attention to a few key readings; it illustrates the degree to which the operation of human and computer experts are converging.

A serious accident in Seveso (Italy), in which 2 kg of the severe poison dioxin was released into the atmosphere, was believed to be due to a runaway chemical reaction. Exothermic reactions gradually increased the temperature of material in a reactor which had been incorrectly shut down. The increased temperature promoted further exothermic reaction, which led eventually to explosive rupture of a safety disc and loss of much of the reactor's contents. In the moments immediately before such an explosive outbreak, the temperature may be rising at the rate of tens of degrees per second. In such circumstances, making the right decision, and making it quickly, is obviously crucial.

Adapting the knowledge base in real time

Knowledge bases in advisory expert systems are largely static. The knowledge base of a system working in tenancy law might consist of information on relevant legislation and recent legal cases, and, once built, will change only slowly, possibly remaining unaltered for months at a time.

By contrast, the knowledge base of systems controlling equipment or industrial plant is fluid, and contains both fixed and variable knowledge. Data may change substantially within minutes or seconds, and there may be hundreds or thousands of readings that the expert system must monitor.

Expert systems which retain dynamic data within the knowledge base must have some means to represent facts like 'the green filter is in the light beam'. Furthermore, the inference engine must be able to bring about physical changes and adapt the knowledge base to reflect these changes.

A real-time expert system has a complete record of the present state of items over which it has control in the form of entries in its **world list**:

W = {IN(RED_FILTER,SAMPLE_BEAM),
 AT(MAIN_GRATING,540), ...}

This indicates that the red filter is in the sample beam, that the grating is positioned to pass light with a wavelength of 540 nm and so on.

When the expert system decides that a change must be made, actions are requested by the inference engine executing an instruction such as:

MOVE[RED_FILTER, SAMPLE_BEAM, REFERENCE_BEAM]

This has a twofold effect. First, a message is sent to the spectrometer to bring about a physical change (in this instance, the movement of the filter from one position to another). Secondly, changes are made to the knowledge base by making deletions from and additions to the world list. If the filter is moved from sample to reference beam, this change can be recorded by deleting

IN(RED_FILTER,SAMPLE_BEAM)

from the world list, and substituting

IN(RED_FILTER,REFERENCE_BEAM).

However, while it may be simple for the inference engine to ask for a change, it may not be so straightforward to actually bring that change about. There are situations in which a task can be completed only by first taking a step in the wrong direction; the specialized reasoning of the expert system can then be invaluable.

Suppose an expert system controls the issue of items from a scientific hardware store. The store contains boxed equipment to be retrieved by a robot storekeeper, which searches through the store under the direction of the expert system and collects the requested items for delivery.

In the interest of efficient storage, boxes may be piled several high on each shelf and an item which the robot must retrieve may be at the bottom of a stack. The expert system must be aware that to achieve its goal (retrieve a certain box), it may need to tell the robot to do something which initially is counter–productive: boxes may have to be moved from the shelves where they belong to adjoining shelves, where they do not, to get at the box required. The first step that the expert system must decide upon is thus to put things in the wrong place (Fig. 3.9).

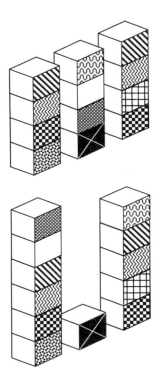

Fig. 3.9 An example of a task in which the first step moves the system away from its goal.

This is a trivial example, but in an industrial plant related problems on a larger scale may require an expert system to display a high degree of intelligence, so that every operation is carried out in a safe, logical, and yet direct way.

Similar considerations apply to scientific instruments. Some instruments contain few movable components, and relationships written into the knowledge base can then define every conceivable safe move. If there are dozens of components, however, there may be thousands of possible moves, and the knowledge base might not be able to accommodate instructions for every one. (Indeed, instructions for some moves might be unknown.) The expert system then comes into its own, determining a safe path for every move before starting.

3.9 Limitations of expert systems

It is not difficult to identify situations in which an expert system might be useful. If a task requires the advice of a human expert, a computerized expert can, in principle, help. However, in practice things may not be so straightforward.

It is essential that suitable experts be available and willing to provide input into the system; if none exist the system clearly cannot be built, no matter what its potential.

A more fundamental difficulty is that it may not be possible to re-cast expert knowledge into a form that can be fed into the knowledge base (recall the 'fluffy precipitate'). Both the data and the task confronting the expert system must be capable of being suitably coded.

IUPAC: International Union for Pure and Applied Chemistry.

There are tasks that a human expert could complete, but that a computerized expert would find almost impossible. Suppose IUPAC chose to devise a new method for the systematic naming of inorganic compounds; it would be essential to enlist the help of experts in inorganic chemistry. However, an expert system could contribute little, even if it knew the present names of all inorganic chemicals, since such a systematic re-naming requires not just the application of rules to a problem, but the creative effort of making new rules, and the very personal judgement of whether those rules are 'clear', 'logical', 'understandable' and so on. In this kind of application expert systems are impotent.

Tasks also exist which are inherently uncertain because they are based on intangible data. The selection of a candidate by interview is one example. 'I know she's just the person – she made such a good impression, don't you think?' Could such a judgement be made by a computer? And if it were, would we allow the computer to make the final choice if we preferred a different candidate? Selection of candidates by interview is something of an art (the unsuccessful candidates might call it a lottery); the rules defining it are almost certainly so vague as to prevent the building of an expert system that could take over from humans.

Progress is being made in the development of expert systems which can make subjective assessments. By associating human evaluations of texture, fluidity, sun-screening power, and aroma with the constituents of sun blockers, a major pharmaceutical company in Britain is using an expert system to formulate sun-blocking products. Personal judgement of the smell and texture of the product plays a significant part, but the expert system is successfully devising new products of high quality by combining human knowledge with its own logical abilities.

3.10 The application of expert systems to chemistry

In view of the limitations discussed above, you might be wondering if, after all, there is any place for expert systems in science. There are many applications in which expert systems are potentially valuable, but it is important to be aware of these limitations. The key question to be answered when considering the construction of a new system is whether the task and knowledge can be coded in a suitable form.

This may be tricky; experienced designers can look at a prototype car on a computer screen and know its shape is 'not quite right', but be unable to explain what is wrong. Chemists formulating new perfumes meet similar difficulties. To know what the public will think is a desirable perfume, or an attractive shape for a car requires an expert, but transforming an expert's opinion on such matters into rules that can be placed into a knowledge base is difficult. It is not easy for an expert system to make such subjective assessments reliably.

In science the scope for expert systems is substantial, but the task that they face is significant. A scientific expert system should be able to cope with error-loaded or incomplete data, do everything a human expert can do in a limited domain, and be able to explain its reasoning. This is quite a shopping list, and suggests that a considerable amount of work might be needed to develop a good expert system.

To reduce this task to manageable proportions, expert systems are usually constructed from a commercially-available 'shell'. This provides the basic components of the system – an empty knowledge base, an inference engine, and a user interface – as a package, so that the user can concentrate on building a stock of assertions and rules in the knowledge base, the core of a productive working system. Use of a shell also reduces the effort required for maintenance, as the knowledge base expands.

Expert systems built in this way are now helping scientists in molecular modelling, chemical kinetics, analytical methods development, organic synthesis, imaging, diagnostic systems, and many other areas. Their most widespread use in chemistry is presently in the control of equipment and robot samplers, and acting as advisers in analytical procedures.

In the analytical laboratory a good system might:

- advise the user on methods of analysis for different samples;
- interpret spectra, chromatograms, electrochemical measurements, and other data;
- instruct novice users of the system;
- monitor equipment performance;
- report defects, and offer help when errors are detected;
- schedule work and instruments if samples must be processed with different priorities;
- control instruments, robotic samplers or process lines.

It will be clear that any system which can meet all these objectives will have a significant part to play in the operation of a large laboratory.

The value of an expert system depends markedly on the quality of information within it and the sophistication of the shell with which it was built. As growing computer power encourages the development of wider knowledge bases, the growth and diversification of these systems in the laboratory is certain to continue.

Further reading

Jackson, P. (1990). *Introduction to expert systems*. Addison-Wesley, Wokingham.

Lucas, P. and van der Gaag, L. (1991) *Principles of expert systems*. Addison-Wesley, Wokingham.

4 Genetic algorithms

4.1 Introduction

It can be difficult and frustrating trying to solve scientific problems. The difficulties arise for a variety of reasons. Some problems are so obscure that it seems impossible to discover any answer to them at all. Other problems present just the opposite difficulty: they offer us a huge over-supply of answers, within which the best is hidden under a multitude of inferior competitors.

Many scientific problems, such as the determination of the conformation of macromolecules, fall in this second category. Often there are several different approaches available to tackle these problems; among the most productive are techniques capable of sophisticated searching.

The genetic algorithm (GA) is just such a technique – an intelligent way to search for the optimum solution to a problem hidden in a wealth of poorer ones.

The genetic algorithm is an optimization technique based on evolutionary principles.

Evolutionary ideas and terminology pervade the GA. The algorithm works with a 'population' of individuals, each of which is a candidate solution to the problem; these individuals 'mate' with each other, 'mutate', and 'reproduce' and in this way evolve through successive generations towards an optimum solution.

It is curious that evolution should provide the inspiration for solving numerical problems, and remarkable that it forms the basis for a method of great power and versatility. In this chapter we shall learn what characteristics of evolution give it this ability, and how its power can be harnessed in a most effective and intriguing way.

4.2 What can the GA do?

At first sight problems in science that involve searching seem rare in comparison with those that require other kinds of computation, such as quantum mechanical or statistical calculations. However, this impression is misleading. Searching problems appear uncommon only because they are usually presented in a form that obscures the fact that very many solutions to them actually exist.

These problems span a wide range; for example:

● *How can one calculate the geometry that a molecule will adopt to minimize its free energy?* The energy of interaction between neighbouring groups in a molecule depends on the interatomic dihedral (torsional) angles and distances between the groups (Fig. 4.1), so different conformers (arrangements of atoms in space) have different total energies. The techniques of quantum mechanics can in principle identify the optimum conformations, but for molecules of the size of proteins these are very large-scale calculations indeed, and AI searching algorithms may be superior to quantum mechanical methods.

Of the infinite number of conformers possible for a large molecule, only one or a few will have the minimum energy.

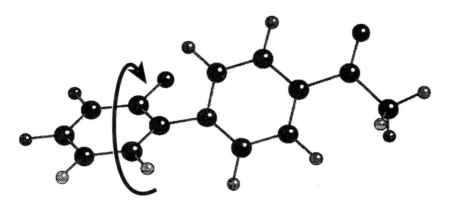

Fig. 4.1 The energy of a molecule depends upon its torsional angles.

● *A UV/visible absorption spectrum is the sum of band spectra that arise from different electronic transitions; how can the underlying spectra be found?* The number of bands composing a UV/visible spectrum may be small, but there are many ways to decompose the spectrum, depending upon the assumptions made about the number and shape of the underlying bands. These assumptions may rest on a rather flimsy theoretical base, and if they are incorrect, it may prove impossible to satisfactorily resolve the spectrum.

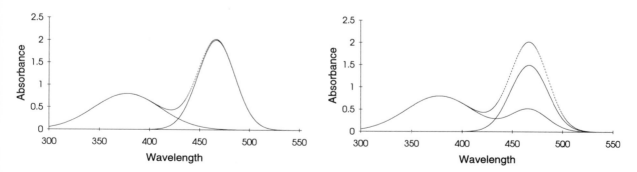

Fig. 4.2 A UV/visible spectrum decomposed in two different ways into possible underlying spectra.

● *How many isomers of $C_{26}H_{52}O$ are ethers containing just one six-membered ring?* In principle we can determine how many isomers contain certain functional groups by drawing the structure of each isomer and checking to see whether it contains the features of interest. However, the number of isomers sharing this formula is very large and only a tiny fraction of these are ethers containing a six-membered ring, so that in practice this procedure is impossibly time-consuming.

At the heart of each of these problems is the need to single out a few optimum solutions from the large number of possible solutions available; they suggest the diversity and complexity of searching tasks. We recall from Chapter 1 that AI is increasingly the method of choice for solving large-scale, complex problems, so, when problems of this sort arise in science, there should be a role for intelligent algorithms in finding a solution of acceptable quality.

4.3 What makes the genetic algorithm different from other methods?

Any computational task (or chemical task, for that matter), which involves mating, mutation, and reproduction sounds definitely odd, if not illicit, so you will not be surprised to discover that the GA has little in common with most other optimization and search methods. The major differences between it and conventional methods are:

● *The GA is a stochastic algorithm, not a deterministic one.*
Suppose the solutions to a problem lie on the surface shown in Fig. 4.3. A conventional algorithm searching for the optimum solution (the maximum on the surface) will investigate it in a clearly-defined and predictable fashion. For example, in 'steepest-ascent hill-climbing', one of the most widely-used conventional algorithms, an arbitrary starting point is chosen; the gradient at this point is found and used to define which direction is 'up'. A short step 'upwards' is taken and the gradient found at the new location. This procedure is repeated until a maximum is reached; hill-climbing then terminates.

This procedure is **deterministic**, because the behaviour of the hill-climber is fully determined by the rules that govern its operation. The existence of these rules means that we can predict before starting exactly how the algorithm will function. Furthermore, its behaviour is completely invariant, so every climb of the same surface starting from the same point follows exactly the same route. For very simple surfaces hill-climbing is fast and reliable, but it encounters serious and sometimes fatal difficulties when confronted with surfaces containing multiple maxima, or significant noise.

In the former case (Fig. 4.4) there may be too many hills to climb. Where should the search begin? And, once it has found the top of a hill, how does it know there is not a slightly higher hill lurking close by?

Fig. 4.3 How hill-climbing finds the maximum on a surface.

It should be clear that hill-climbing will often find a local maximum, rather than the global maximum, on a surface.

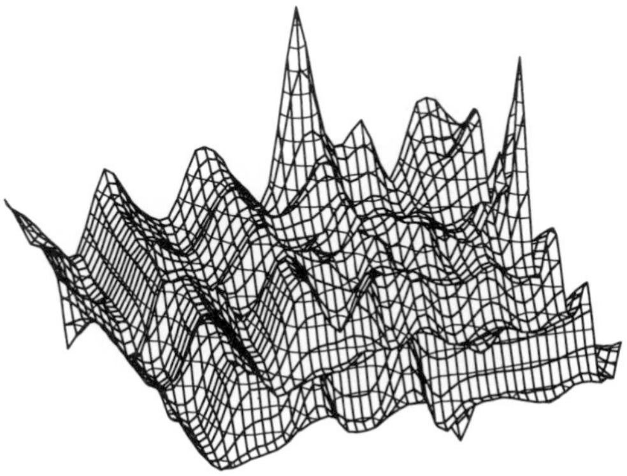

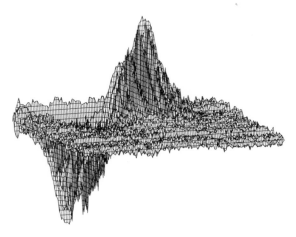

Fig. 4.4 A surface with numerous maxima.

Fig 4.4 shows a small portion of a complicated surface, derived from studies of flowshop scheduling. There are roughly 10^{15} maxima in all.

In Fig. 4.5 the noise makes it hard for a hill-climber to know which direction is 'up'; it may rapidly become trapped by a peak consisting of more noise than signal.

Fig 4.5 shows experimental data derived from studies of kinetics by electron spin resonance.

Fig. 4.5 A surface containing substantial noise.

By contrast, the GA is **stochastic**. A stochastic process relies upon random elements ('chance') in parts of its operation. This is illustrated by Fig. 4.6, which shows two GA searches of the same surface. Both searches reach the maximum, but, though they started from the same point, they pursue quite different routes, since at each point visited random factors influence the decision that the algorithm takes on where to move next.

On the face of it this seems to be an inefficient way to search; if there is a best way to climb the surface, why not use it? The difficulty is that the 'best' route is usually unknown; hill-climbing presumes that the fastest way to the top is via the steepest hill, but although this is one method of locating hilltops, it is not the only one. Any deterministic method such as hill-climbing is certain to meet surfaces which defeat its rules, and it will then be inexorably

Fig. 4.6 Routes that might be taken by a GA searching the surface shown in Fig. 4.3.

lured into the same trap on the surface every time it is run (Fig. 4.7). A randomized method like the GA is less likely to be misled by such traps, and if the search stumbles into one, a mechanism exists whereby it can escape and the search be resumed elsewhere.

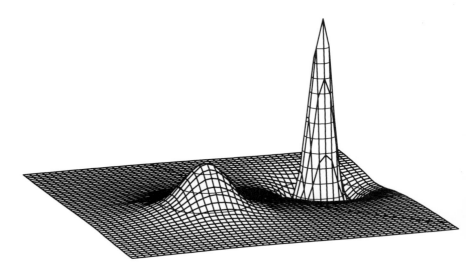

Fig. 4.7 A surface which would defeat a hill-climber.

● *The GA investigates many possible solutions simultaneously; each investigation learns about a different region of the surface.*

It is as though numerous hill-climbers were wandering around on the surface at the same time, each one having started from a different, randomly-chosen position.

This is reminiscent of the communal memory in an artificial neural network. However, though an application requires just one neural network, in the GA the whole population works co-operatively, and all individuals together constitute the memory. There must be value in numbers, and we shall shortly see why this is so.

It may seem counter-productive to spend time investigating many solutions if there is only a single right one, or a few, but these multiple GA solutions are not independent of one another; instead the GA treats all solutions as a group. This group finds better answers than a single search on its own can, and finds them more quickly, because each solution can communicate with others in the group (as we shall see, this is where the 'mating' comes in). When one solution begins to move towards a good solution, information about this improvement is disseminated through the rest of the population, and other solutions use and benefit from this new knowledge. In effect a collective memory develops, spread among many solutions. Communication and interaction of this sort, the nature of which will become clear shortly, are essential features of the GA and are fundamental to its success.

● *The GA requires no auxiliary information about a surface, such as the gradient at a point.*

This is a crucial advantage, which greatly extends the range of problems that the GA can tackle. Calculus-based methods depend upon the existence of derivatives, and numerical hill-climbers require that the gradient at any point gives reliable information about which direction is 'up'.

This is often not the case in noisy scientific data, and hill-climbers have severe difficulty travelling across surfaces such as that shown in Fig. 4.5. By contrast, the GA is little troubled by the presence of even substantial noise, and completely unaffected by the absence of derivatives.

4.4 The genetic algorithm: the mechanics

Although the GA is based upon the concepts of evolution and is full of evolutionary terminology, evolution-based methods are logical, not biological, in nature. Every individual in a GA **population** is a distinct *numerical* solution to the problem; these individuals are subjected to evolution-like operations, but no understanding of biology is necessary to understand how these processes work. It is these evolutionary manipulations – the mechanics of the GA – that we now consider.

The GA is a cyclic process, in which a sequence of operations is executed repeatedly in an attempt to drive the search toward optimum solutions. Each cycle, in which a population of solutions is first assessed for fitness, then reproduced and adapted, constitutes a **generation** and consists of the following steps:

The genetic algorithm

1. On the first cycle only, form a starting population of **chromosomes** or **strings**. These are candidate solutions to the problem.
2. Determine how good a solution each string is. Stop if a high-quality solution exists, or the maximum number of generations has passed.
3. Determine the **fitness** of each string from its quality.
4. Use a **reproduction operator** to form a new population by selecting strings from the current population with a probability determined by their fitness.
5. Choose pairs of strings at random and combine them using a cut-and-pasting procedure defined by a **mating** or **crossover operator**.
6. Alter a few members of the population using a **mutation operator**.
7. Return to step 2.

A typical GA calculation requires hundreds or thousands of these cycles to evolve good answers.

It is not obvious that the procedure outlined in the box is capable of doing anything useful at all, let alone solving complex scientific problems. Nor is it clear why this set of steps deserves to be called 'intelligent'. To appreciate these points, we will consider each step in more detail, illustrating the operation of the GA as we do so by applying it to a chemical problem.

Prepare the initial population

A string is an ordered sequence of numbers which in some way represents a solution to a problem; the terms 'string' and 'chromosome' are used interchangeably in GA work. By analogy with natural systems, each position within a string is known as a **gene**, and its value, an **allele**.

Suppose we wished to use the GA to find the optimum conformation of the fluoroalkane shown in Fig. 4.8(a), 3,5-difluoroheptane.

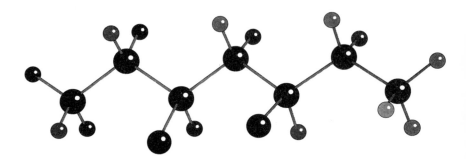

Fig. 4.8(a) We will use the GA to find the optimum conformation of this molecule.

The bond angles at each carbon atom are tetrahedral, and the bond lengths are largely independent of stereochemistry, so one conformer is distinguished from another only by the values of its dihedral angles. Any conformer can be identified uniquely by specifying these dihedral angles in order $\{\alpha_1, \alpha_2, \alpha_3, \alpha_4, \alpha_5, \alpha_6\}$ so that $\{183, 165, 313, 253, 294, 84\}$ for example, represents the conformer shown in Fig. 4.8(b). This set of numbers constitutes a string, and it is numerical representations of this sort that will be manipulated by the operators within the algorithm.

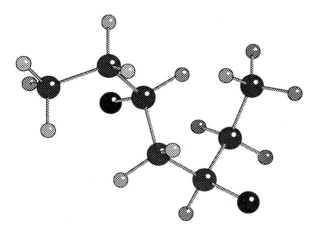

Fig. 4.8(b) A randomly-chosen conformer of the fluoroheptane.

Each angle can take any value within the range 0 to 359° (0° and 360° are, of course, equivalent), so there is strictly an infinite number of possible conformations. (There are about 10^{15} conformations if, as in this example calculation, we restrict the angles to integer values; there is also some degeneracy among solutions, but 10^{15} is still rather a large number of solutions to look through to find the best ones.)

It is the task of the GA, starting from a number of random conformations of this sort, to transform these into good – and ultimately optimum – conformations. In pursing this goal, the GA uses several simple operators to manipulate strings. The first step in the calculation is to form an initial population of strings, by selecting random conformers. We must therefore decide how many strings constitute a population.

As generation succeeds generation, the angles that make up the strings change as they are manipulated by the algorithm; the strings gradually evolve in this way towards good solutions. Just as in biological systems, fruitful evolution requires variety in the population, and GA experiments show that populations containing just a handful of strings cannot provide the diversity on which evolution thrives; this effectively provides a lower limit to the population size.

On the other hand, though a very large population should be diverse, good strings may then find themselves overwhelmed by a flood of marginally poorer strings, and the calculation will be slow to converge. This suggests that there is also an upper limit on population size. Bearing in mind these restrictions, populations containing 25–100 strings are usual in GA work.

For the fluoroalkane calculation we will use a population of slightly smaller size than is normal, since this will adequately illustrate the mechanics of the algorithm. As we shall see, even this small population will rapidly discover near-optimum conformations.

We form the members of the first generation by choosing at random six angles in the range $0 \leq \alpha \leq 359$ for each conformer in a total population of ten. These starting strings are shown in Table 4.1.

The goal of the GA is to find a solution in which all dihedral angles are at, or close to, their optimum values, at which the interaction energy between the groups separated by those angles is at a minimum. These energies (in arbitrary units) depend upon the dihedral angles between the groups in the fashion depicted in Fig. 4.9. Angles in the shaded regions correspond to an interaction energy within 2 per cent of the minimum. Figure 4.9 shows that a 'good' angle between a methyl group and a methylene group is one in the range 55°–65°, 175°–185° or 295°–305°, and for the CHF–CH_2 interaction 'good' angles are those between 175° and 185° inclusive. In Table 4.1 the good angles defined in this way in the first population have been shown in bold face. As we would expect, since the angles that constitute the strings in this population were chosen at random, there are few good angles at this stage.

The problem described here is related to the important 'protein-folding problem'. The development of new drugs relies increasingly on quantum mechanical studies of the strength with which a potential drug binds to a target protein molecule. Such studies can help narrow the range of molecules whose development might be worth pursuing, and thus control the costs of their synthesis and laboratory trial. The calculations depend for success on a reliable knowledge of protein stereochemistry, in particular in the region of the active site. The structure of proteins is difficult to determine experimentally, so extensive efforts are being made to calculate how a protein folds to minimize its free energy, and thus determine the stereochemistry around the active site. This calculation is of great complexity, but is of enormous academic and commercial interest.

Table 4.1 The strings in generation 1.

String	α_1	α_2	α_3	α_4	α_5	α_6	Energy	Fitness
1	**183**	165	313	253	294	84	6.502	0.180
2	340	46	8	5	36	98	15.374	0.069
3	**302**	105	80	45	137	97	9.954	0.111
4	318	232	324	302	249	119	12.616	0.086
5	**300**	297	72	318	344	196	7.040	0.164
6	72	327	355	20	330	143	14.900	0.072
7	342	231	68	68	325	141	11.770	0.092
8	287	203	7	78	142	76	9.338	0.119
9	203	103	343	70	**180**	51	8.308	0.136
10	125	227	145	114	336	117	17.438	0.061

Best fitness 0.180 Average fitness 0.109

Determine the quality of the strings

Once a starting population has been chosen, each string must be inspected to determine its quality. Since the most stable conformer is that of lowest interaction energy, a high-quality solution to this problem is one of low energy.

The 'quality' of a string is simply a measure of how good a solution it is.

The total interaction energy for a given string is the sum of the energies arising from interactions between neighbouring –CHF– and –CH$_2$– groups, and between neighbouring –CH$_2$– and –CH$_3$ groups along the carbon backbone of the molecule. Although interactions between non-neighbouring groups are not zero, they are smaller than interactions between neighbours, and we shall ignore them. This is only a minor approximation and makes no difference to either the principles of the calculation or its success.

We can use Fig. 4.9 to estimate the total interaction energy for a conformer by summing the energies associated with each pairwise interaction. For example, for string 1 the energy associated with dihedral angle α_1 is 0.19, that with dihedral angle α_2 is 0.66, and so on. The total energy is then given by:

Energy

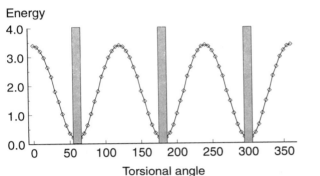

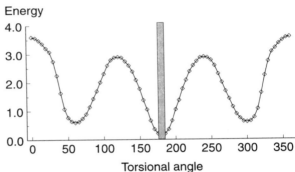

Torsional angle

Fig. 4.9 The dependence of the interaction energy upon dihedral angle for the methyl-methylene (left) and the methylene-fluoromethylene (right) interactions.

$$E_i = 0.19 + 0.66 + 0.95 + 2.68 + 0.65 + 1.37 = 6.50$$

In this way the total energy for each string in the starting population can be calculated, and these are shown in Table 4.1.

Determination of the quality of each solution is the only point of contact between the algorithm itself and the physical problem that it is set to solve – it is the only kind of information that the GA needs. Most searching algorithms require access to extra information. For example, numerical hill-climbers need to know the gradient of the surface at every point that they visit so that they can decide in which direction to take their next step. This extra information may be missing (on a fragmented surface, for example), or it may be misleading (on a noisy surface). Even when the information is available, time is required to find or to calculate it.

By contrast, the GA is not concerned with the nature of the immediate (or distant) environment around a point; it needs to know only how 'good' a solution that point represents. In principle this information is always available, (the problem would be insoluble if a good solution could not be distinguished from a bad one), so the GA can in theory search a great variety of surfaces.

That the GA has such a limited appetite for knowledge about the problem is an important advantage over other methods. It makes the GA a very general method of analysis indeed, and in this respect it has much in common with neural networks. Just as a single neural network can be trained to solve any one of numerous different types of problem, a single GA program can be adapted to undertake a wide range of tasks through minor changes in computer code.

In performing this calculation we must remember that α_1 and α_6 refer to methyl–methylene interactions, and that all other angles refer to methylene–fluoromethylene interactions.

Determine the fitness of the strings

This is the first appearance in the GA of a factor which drives the algorithm towards optimum solutions.

Evolution embodies the principle of 'survival of the fittest': 'fit' organisms in one generation generally manage to leave behind progeny for the next, while 'unfit' ones often do not. In GA terms, unfit strings are those representing poor solutions, and their continued presence in the population will hinder the calculation. Under a survival of the fittest scheme, unfit strings usually fail to live for more than a few generations beyond their first appearance, as fit strings take their place. Evolution-like manipulations in this way massage the population and encourage the growth of good solutions, as inferior ones are weeded out.

The selection of an appropriate relationship between quality and fitness becomes easier with experience, and in simple problems, such as the present one, choice of a good relationship is in any case not difficult.

Survival of the fittest is a key element of the GA, and to put it into practice we must derive a relationship between the quality of a string, which we already know, and its fitness, which we require. This is an important step, since the rate at which the algorithm converges will be affected by the relationship chosen.

The relationship must suitably reward good solutions with high fitness if these solutions are to thrive; beyond that criterion, the relationship is essentially arbitrary. It is not hidden away somewhere, waiting for us to discover it, nor is it predefined by the constraints of the problem; instead, any recipe can be chosen. In this problem fitness is inversely related to quality (as the energy decreases, the strings get better), and we shall choose to relate f_i, the fitness of string i, to E_i, its energy using the function:

$$f_i = 1.0/(E_i - 0.9) \tag{4.1}$$

(The purpose of the 0.9 in eqn (4.1) is to accentuate the difference in fitness between good and poor quality strings. If fitness were defined simply as the inverse of the energy, the best possible string would have a fitness of 1.042 since the minimum energy is 0.96; the worst possible string would have a fitness of 0.047. Using eqn (4.1), the maximum fitness is 16.667, and the minimum fitness is 0.049. The increased spread of fitness that the introduction of the 0.9 in eqn (4.1) provides has the effect of putting extra evolutionary pressure on the poor strings; this will result in them being removed from the population more effectively, which in turn speeds convergence.)

The fitnesses of strings in the first generation calculated through use of eqn (4.1) are shown in the final column of Table 4.1

The first population appeared through a sort of immaculate conception, but subsequently each fresh population must be derived from the previous one.

Which strings shall reproduce?

The GA works in a series of generations, in each of which a population is formed, then allowed to reproduce. The first step in this process is the selection of parent strings for reproduction.

In this, we follow evolution rather closely. Fit organisms are generally successful in transmitting their genes from one generation to the next, while unfit ones usually are not. It was with this in mind that we calculated the fitness for every string; parents are now selected from the current generation on the basis of that fitness.

Survival of the fittest provides guidelines by which to construct the next generation, but it is not a completely deterministic recipe: sometimes poorly-adapted individuals manage to leave offspring, despite their unpromising fitness; sometimes fit individuals fail to do so. So, it would not be appropriate to uncritically choose for the next generation every string with high fitness and disregard all those with low fitness. A method of selection is needed which, while biased in favour of highly fit strings, still gives unfit strings some chance of reproducing.

Fig. 4.10 The roulette wheel, which functions as a reproduction operator.

The roulette wheel

A simple way to accomplish this is to use a roulette wheel. We construct an imaginary roulette wheel in which each string is allocated a slot with a width proportional to the string's fitness (Fig. 4.10). The wheel is spun, and the string into whose slot the imaginary ball falls is copied once into the new generation. This process is repeated until the number of copies made equals the population size.

The roulette wheel mechanism clearly meets both the requirement that fit strings have the best chance of reproduction (since they have the widest slots), and that unfit strings have a lesser but non-zero chance of being copied. A string of exactly average fitness has an even chance of being copied once into the next generation, so we can calculate the number of copies of each string that the roulette wheel should generate by scaling the fitnesses so that the average is 1.0. The scaled fitness then equals the expected number of copies of that string in the new generation. Table 4.2 shows these scaled fitnesses, and also shows the actual number of copies that the roulette wheel mechanism generated for our illustrative calculation (which, of course, must be a whole number for each string).

The set of child strings shown in Table 4.2 forms the starting point for generation 2; each string is a copy of a single parent in the first generation, two copies of string 1, one each of strings 2-9, and none of string 10. A new population generated by the roulette wheel is almost certain to contain multiple copies of some of the fitter strings, and be without copies of some of the less fit strings, because of the bias of the selection process towards the fitter strings.

This bias is not marked because the difference in fitness between the best and worst strings is as yet quite small; as a result, the roulette wheel shows little discrimination in making the new population: poor strings and good strings are both likely to be selected. However, the discrimination increases as the calculation proceeds, and the difference in fitness between good and bad strings grows. To anticipate later results, although in population 1 the range of unscaled fitnesses is 0.0523 to 0.0227, by population 20 the range is 3.125 to 0.344, and in population 97, the range is from 9.09 to 0.299. By then, the roulette wheel is more than 30 times as likely to reproduce the best string as it

Table 4.2 Scaled fitnesses and number of copies made by the roulette wheel.

String	Fitness	Scaled fitness	Copies
1	0.180	1.650	2
2	0.069	0.636	1
3	0.111	1.018	1
4	0.086	0.789	1
5	0.164	1.505	1
6	0.072	0.661	1
7	0.092	0.844	1
8	0.119	1.092	1
9	0.136	1.248	1
10	0.061	0.560	0

is to reproduce the worst, and survival-of-the-fittest clearly has a dominant effect on the composition of the new population.

A large range of fitness is broadly beneficial, since it encourages the proliferation of good solutions, and the discarding of poor solutions. Within a few generations of the start of the calculation, as the range of fitness grows, the best strings are being reliably reproduced from one generation to the next, while poor strings are being discarded with similar efficiency.

Because the roulette wheel is biased towards fitter strings, the average fitness of the new population will generally exceed that of the parent population. This is progress of a sort, but in reality not much has been accomplished yet. There are no solutions in the new population that were not present in the first one, and we have yet to improve upon any of these initial, random, solutions. We need to find some method by which new high-quality strings can be generated.

Swapping genes: the crossover operator

The most fundamental role in the creation of good solutions is played by the crossover operator; this performs the GA equivalent of evolutionary mating.

The mixing of genes that accompanies sexual reproduction ensures that biological offspring have characteristics derived from both parents. A particularly fortuitous combination of genes may lead to a child significantly better adapted to its environment than either of its parents – a fitter individual. Because of the enhanced fitness that this child enjoys, it should in turn contribute its own descendants to the next generation, so that the favourable combination of genes will survive and, indeed, proliferate through succeeding generations, as less fit competitors are elbowed out of the population.

The multiplication of fitter individuals at the expense of the less fit drives biological evolution; a parallel process pushes the GA towards good solutions.

Two GA strings mate through the good offices of the crossover operator. This cuts a pair of randomly-chosen strings at some position and swaps the cut segments (Fig. 4.11).

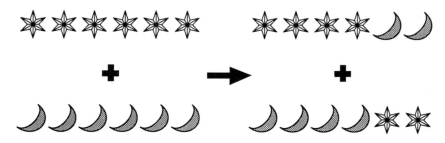

Fig. 4.11 One-point crossover.

The crossover operator is applied to pairs of child strings chosen at random from those selected by the roulette wheel. In our example, child strings 1 and 3, which are

{183,165, 313, 253, 294, 84} and {302, 105, 80, 45, 137, 97},

are chosen for crossover from the new population. A cut is made arbitrarily between positions 3 and 4, and the cut segments swapped to yield the new strings

{183, 165, 313, 45, 137, 97} and {302, 105, 80, 253, 294, 84}

This is **one-point crossover**, in which a single cut is made; **two-point crossover** (Fig. 4.12) is also widely-used (though we shall not use it in our example), and other crossover operators may be used in special situations when simple one- or two-point crossover might yield invalid strings, or are inappropriate for some other reason.

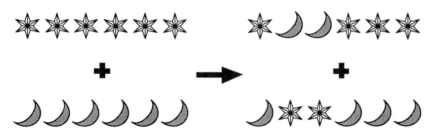

Two-point crossover between positions 2 and 4 of the two strings shown in the text would yield the two new strings:

{183, 105, 80, 45, 294, 84}
and
{302, 165, 313, 253, 137, 97}.

Fig. 4.12 Two-point crossover.

The cut-and-pasting performed by crossover has the effect of shuffling string segments so that strings in the new population inherit characteristics from two members of the old one. If both parents were fit strings, then by analogy with biological systems, one or both of the offspring stand a good chance of being fitter than either parent.

Inheritance by a child string of high-quality segments from two parents is a second factor encouraging the development of high-quality strings in the population.

Crossover is used on most (sometimes all) members of the new population, and in each case the partners to be crossed and the position of the crossover cut are selected randomly. The two strings that result from crossover take the places in the population of the strings from which they were derived. Crossover is not simply a mechanical attempt to reproduce another of the steps of evolution, but, as we shall see, is a powerful way of increasing the overall fitness of the population.

Mutation

In nature, mutations to chromosomes sometimes occur. Strings in the GA may also mutate; this is done by making a random change at an arbitrarily-chosen position in a string. When a string in the sample calculation is mutated, a random angle is replaced by a randomly-selected new one,

In biological systems most mutations are harmful or inconsequential; only rarely does a random mutation yield a significantly fitter individual. However, the population size in natural systems is usually large enough that substantial diversity across the population already exists; biological mating tends to reinforce that diversity and minimizes the chances of stagnation. In the GA by contrast, the population is small, and very vulnerable to loss of diversity; the mutation operator then has an important role to play in preventing the population from becoming too homogeneous.

within the permissible range of 0 to 359°. For reasons discussed in Section 4.6, mutation is applied only infrequently, and in our calculation it occurs at the rate of one mutation per generation. Mutation rates are typically around 1 per 1000 genes, though higher rates are often used for relatively simple problems like the present one. The mutation operator working on the new population selects for mutation the string {318, 232, 324, 302, 249, 119}, which had been placed into population 2 by the reproduction operator, but not crossed with another string, and transforms it into {318, 232, 324, 302, 249, 43}.

The main function of mutation is, through the introduction of new genetic material, to prevent the population from stagnating. Nevertheless, increasing homogeneity in the population, which mutation helps to limit, is an inevitable result of the success of the calculation. As the calculation proceeds, unfit strings are gradually discarded by the reproduction operator, and fit strings proliferate. The cut-and-pasting of crossover tends to disrupt strings, and reduce the chance that a single string will take over the population entirely, but the population still progressively loses variety as inbreeding (crossover between almost identical strings) grows.

If this loss of variety becomes severe, convergence slows, and in the absence of mutation the algorithm may become trapped at a sub-optimal solution and be unable to escape. The introduction of random mutations at a small rate continually injects diversity into the population and prevents the reproduction operator from filling the population with large numbers of good, but sub-optimal strings.

After reproduction, crossover, and mutation, the members of generation 2 are as shown in Table 4.3.

The next generation

The new population is now complete, and the algorithm returns to step 2. The interaction energy for each new string is calculated and, from these, the new fitnesses are found; these are shown in Table 4.3. As the table shows, the average fitness has risen, and the best string is better than the best in the first population; better solutions have arisen following the evolution-like manipulations described above.

But we should be cautious about congratulating the algorithm on this improvement: perhaps the higher fitnesses are mere chance. After all, roulette wheel selection preferentially selects high-fitness strings as the starting point for the new generation, so the average fitness should improve, provided that the disruption caused by crossover and mutation is not too great. Furthermore, the population is small, and the initial random population might have been a peculiarly unfit bunch, from which the mix-and-match of crossover just happened to generate a fitter best string. Does the improvement in fitness really arise from the manipulations of the GA itself?

Table 4.3 The strings in generation 2.

String	α_1	α_2	α_3	α_4	α_5	α_6	Energy	Fitness
1	**183**	165	68	68	325	84	5.866	0.203
2	**183**	165	313	45	137	97	7.834	0.154
3	203	103	343	318	344	196	12.540	0.086
4	340	46	8	5	36	76	13.580	0.079
5	**300**	297	72	70	**180**	51	2.808	0.538
6	**302**	105	80	253	294	84	8.622	0.130
7	287	203	7	78	142	98	11.132	0.098
8	318	232	324	302	249	43	10.022	0.110
9	72	327	355	20	330	143	14.900	0.072
10	342	231	313	253	294	141	12.406	0.087

Best fitness 0.538   Average fitness 0.155

The way to find out is to run the algorithm a little longer. After ten generations (Table 4.4) the average fitness is 0.4728, and that of the best string is 0.7825. Even the worst string now has a fitness that exceeds that of the best string in generation 1. The strings bear little resemblance to their ancestors in the first generation – nine cycles of cut-and-pasting by the crossover operator have seen to that – but there can be no doubt that fitter strings are being created. As Fig. 4.13 shows, as generations pass there is a steady move towards good – and eventually optimum – solutions.

Evolution really does solve chemical problems. But how does it do it?

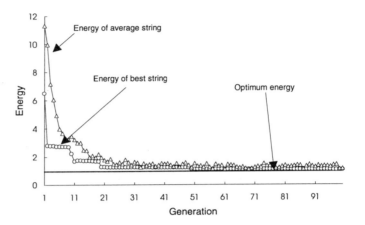

Fig. 4.13 The variation of energy with generation number.

Table 4.4 The strings in generation 10.

String	α_1	α_2	α_3	α_4	α_5	α_6	Energy	Fitness
1	**300**	297	68	70	**180**	51	2.704	0.570
2	**183**	**183**	297	70	**180**	51	2.228	0.783
3	**300**	165	68	70	**180**	51	2.746	0.557
4	**183**	297	78	70	**180**	51	3.078	0.470
5	**300**	297	78	70	114	51	5.750	0.208
6	**300**	297	78	70	**180**	51	3.048	0.477
7	**297**	68	78	70	**180**	51	3.156	0.453
8	**183**	165	78	70	**178**	51	3.140	0.457
9	**183**	297	68	70	**180**	51	2.734	0.561
10	**300**	297	78	70	**180**	240	6.188	0.194

Best fitness 0.783   Average fitness 0.473

Note how much the homogeneity of the population has increased by generation 10, as the algorithm starts to 'home in on' a good answer.

4.5 Why does the GA work?

It is the aim of the GA to derive strings with the highest possible fitness. In this problem the fittest strings are those that contain several 'good' dihedral angles in the sense that we defined earlier.

Strings which contain one or more good angles should have above-average fitness (though, particularly in the early generations, the beneficial effect of these angles will sometimes be cancelled out by the presence of some particularly poor angles in the same string). When the reproduction operator selects strings for the next generation, it preferentially picks high fitness strings, and may make several copies of the best of them. It was the presence of good angles in these strings that made them fit, so *the effect of the reproduction operator is to cause proliferation not just of fit strings, but also of the good angles within them.* This effect is illustrated in Table 4.5 which shows how the number of good angles in the population grows as the algorithm runs.

The crossover operator also promotes the growth of good angles. Consider the following two strings, both of which are members of generation 1:

{**300**, 297, 72, 318, 344, 196}   and   {203, 103, 343, 70, **180**, 51}.

Each contains a single good angle, shown in bold face. Each string was reproduced once at the end of the first generation by the reproduction operator, and the two child strings were then brought together and combined by crossover. The application of this operator, cutting between positions 3 and 4, brings together the two good angles in one new string,

{**300**, 297, 72, 70, **180**, 51}.

This string has a fitness greater than that of either parent because of the presence of two good angles, and this will enhance its ability to survive and proliferate in further generations.

The good string, by virtue of its superior fitness, is likely to be selected by the reproduction operator when the time comes to build population 3; in fact in the calculation used in this illustration, the string was copied three times by the roulette wheel. Whereas in generation 1 there was just a single copy of each of the good angles $\alpha_1 = 300$ and $\alpha_5 = 180$, by generation 3 there are three copies of each. Through crossover, either or both of these angles may be spliced into other strings, raising the fitness of these strings in the process. The combined efforts of the reproduction and crossover operators are compounding the number of good angles in the population, even at this early stage.

The two good angles represent part of an optimum solution. A string containing these two angles has, in effect, learnt a little about what a good solution looks like, and is rewarded for this useful knowledge with high fitness. The high fitness ensures that what has been learnt does not die out, but gets passed into subsequent generations. In this way knowledge is copied

Table 4.5 The proliferation of good angles.

Generation	No. of 'good' angles in the population
1	4
2	5
5	14
10	20
20	39
50	49
100	59

The second string generated by crossover of the two shown in the main text is {203, 103, 343, 318, 344, 196}. This string contains no good angles, so should have low fitness. This is indeed the case, and it fails to be copied by the reproduction operator at the end of generation 2 and is lost from the population.

by reproduction from one generation to the next, and gradually infects other members of the population through crossover. Learning is spreading through the population.

Schemata

Early in the calculation, good schemata, that is schemata which confer high fitness on strings that contain them, will be few in number, and short.

We can think of this pair of good angles as a sort of 'building block'. It is the job of the GA to generate such building blocks and shuffle them among strings, gradually assembling them into solutions of higher and higher quality. Though good building-blocks may be widely dispersed in the population, the evolutionary manipulations of the GA will frequently bring together two or more building blocks in a single string. When this happens, the superior fitness of the string thus created will make it likely that this superior solution is not lost, but passed from one generation to the next until displaced by yet better solutions.

Building blocks are a central part of the GA, and their proliferation is the fundamental mechanism by which the GA assembles good solutions. Indeed, the building blocks are so important that they are given a special name, the **schemata** (singular **schema**). Although much of GA theory can be understood in non-mathematical terms, one equation which describes how schemata proliferate in a population occupies a central position in GA work, and an understanding of its derivation will help to clarify why the GA is such a successful searching algorithm.

4.6 The schema theorem

A schema is a **pattern-matching template**. This is a string in which certain fixed positions contain numbers, while all other positions are occupied by an asterisk, which is a 'don't care' symbol. An example of a schema for the conformer problem is {300, *, *, *, 180, *}. A string is said to 'contain a given schema' if the numbers in all the fixed positions in the schema match exactly the numbers in the equivalent positions in the string.

Values in the string in positions in which the schema has an asterisk are of no consequence. For example, the strings {300, 297, 72, 70, 180, 51} and {300, 105, 78, 70, 180, 307} contain the schema shown above, but the strings {297, 300, 70, 72, 180, 51} and {300, 297, 72, 180, 70, 51} do not (Fig. 4.14).

Schemata multiply under the influence of the reproduction and crossover operators as the GA runs; the **schema theorem** predicts the rate at which this will occur.

Reproduction copies strings with a probability proportional to their fitness. Each time the roulette wheel is spun, the probability that string i is copied by the reproduction operator is $f_i/\sum_i f_i$. In a population of n strings, the operator

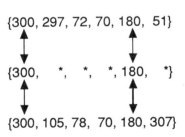

Fig. 4.14 Two strings with a schema in common.

will be called n times per generation and the expected number of copies of string i in the new population will therefore be

$$m_i(t+1) = n f_i / \sum_i f_i \qquad (4.2)$$

You may recognize that $m_i(t+1)$ is the scaled fitness of string i.

in which $m_i(t+1)$ is the number of copies of string i in generation $t+1$ if the string has fitness f_i in generation t.

If a schema H were present only in string i, this equation would tell us how many copies of the schema should occur in the new population. However, one schema may be contained in several different strings, each with its own fitness. The expected number of copies of that schema in the next generation is thus given by

$$m_i(H, t+1) = m(H, t) . n . f_i(H) / \sum_i f_i \qquad (4.3)$$

in which $f_i(H)$ is the average fitness of all strings containing the schema. Making the replacement

$$\bar{f} = \sum_i f_i / n \qquad (4.4)$$

we can rewrite eqn (3) as:

$$m_i(H, t+1) = m_i(H, t) . f_i(H) / \bar{f} \qquad (4.5)$$

Equation (4.5) defines mathematically how a fit schema will multiply, but it also tells us that a schema which confers a fitness *disadvantage* to strings that contain it will tend to be removed from the population, since the ratio of the average string fitness of strings that contain the schema to the average population fitness will be less than one.

The rate at which a good schema proliferates in the population therefore depends upon the ratio of the average fitness of the strings which contain the schema to the average fitness of the population as a whole. Thus, a schema which confers a large fitness benefit upon any string that contains it will multiply rapidly in the population.

It can be shown that, if a schema H is always fitter than the average fitness by a fixed proportion β, then the number of examples of the schema in the population grows according to the equation:

$$m(H, t) = m(H, 0) . (1 + \beta)^t \qquad (4.6)$$

This indicates that the number of copies of a consistently above-average schema rises geometrically in a population, while below average schemata ($\beta < 0$) will be removed from the population at a geometrically diminishing rate. This growth in numbers of high-quality schemata *takes place for all schemata simultaneously*. In any population of significant size, there will be many beneficial schemata, and they will *all* grow in number according to eqn (4.6), as long as the fitness of the strings that contain them remains above the average fitness. It can in fact be shown that the number of schemata processed usefully each generation is of the order of n^3 for a population of n strings. Since typical GA populations consist of between 25 and 100 strings, the number of schemata processed per generation is very large, and this has

The importance of implicit parallelism, introduced in this paragraph, to the success of the GA cannot be over-stated. The evolutionary basis of the algorithm gives it the power to search massive surfaces; implicit parallelism explains how it is able to do this in a reasonable time.

the potential to yield rapid improvements in the average fitness of the population. This simultaneous processing of many schemata is known as **implicit parallelism**, and is a crucial reason why the algorithm can successfully search even massive surfaces.

However, this growth of schemata does not take place unimpeded, since all schemata risk being cut by the crossover operator or destroyed by mutation. Our calculation of the rate at which schemata multiply must take these effects into account.

Crossover will not affect all schemata with equal probability. The crossover operator is most likely to damage those schemata in which the distance between the fixed positions is large. For example, the string $\{300, 297, 72, 70, 180, 51\}$ contains (amongst many others) the schemata $H_1 = \{300, *, *, *, 180, *\}$ and $H_2 = \{*, *, *, *, 180, 51\}$. Under crossover, schema H_1 will be disrupted if a cut is made immediately before positions 2, 3, 4, or 5, while schema H_2 will only be disrupted by a cut immediately before position 6. Schemata in which all the fixed positions are close together are clearly more likely to survive crossover than schemata in which the fixed positions are far apart.

To quantify this, we introduce the **defining length** of a schema, written $\delta(H)$, which is the distance between the first and last fixed position; we will also shortly need the **order** of a schema, written $o(H)$, which is the number of fixed positions. Thus the two schema above both have an order of 2: the first has a defining length of 4 and the second a defining length of 1.

Schema H_1 is destroyed by crossover for four different choices of cut position and survives for one choice; the probability that it will survive crossover is thus 1/5. Schema H_2 is destroyed by one choice of crossover position, but survives if the cut is made in any of the remaining four positions; its probability of survival is 4/5. In general, the probability that a schema will survive crossover is

$$p = 1 - \delta(H)/(l-1) \tag{4.7}$$

in which l is the length of the string.

There may be strings in the new population to which crossover is not applied. Obviously all schema in such strings survive the crossover stage in the formation of the new population, so if the probability of crossover is p_c, the chance of a schema surviving is

$$p \geq 1 - p_c\, \delta(H)/(l-1) \tag{4.8}$$

We have introduced a $\geq$ sign into this expression, since, even if the crossover operator splits a schema, there is a chance that the schema might be re-created if its crossover partner has the same values in the fixed positions as the cut portion. For example, a cut of the string $\{300, 297, 72, 70, 180, 51\}$ between position 3 and 4 will destroy the schema $\{300, *, *, *, 180, *\}$, but if crossover takes place with the string $\{107, 4, 11, 213, 180, 72\}$, the schema is re-created in the new string $\{300, 297, 72, 213, 180, 72\}$.

Combining eqn (4.8) with our expression for the rate at which reproduction causes schema to proliferate, we have

$$m_i(H, t+1) \geq m_i(H, t) . f_i(H) . \left[1 - p_c \, \delta(H)/(l-1) \right]/\bar{f} \qquad (4.9)$$

This equation shows that schema of low defining length contained in strings of high fitness are those which will multiply most rapidly.

The final GA operator is mutation. This destroys schemata at a rate which depends upon the probability of mutation per position in a string, p_m, and the number of fixed positions in the schema (a schema is unaffected by mutation in one of its 'don't care' positions). The chance that a single fixed position will survive mutation is $(1-p_m)$. For the schema as a whole to survive, all $o(H)$ fixed positions must individually survive, and the chance that this will happen is $(1-p_m)^{o(H)}$. The term p_m is always small to guarantee that the disruptive effect of mutation does not destroy good schemata at a significant rate, so this can be approximated to $1 - o(H).p_m$. Thus the final expression for the rate at which schemata are reproduced is

$$m_i(H, t+1) \geq m_i(H, t) . f_i(H) . \left[1 - p_c \, \delta(H)/(l-1) - o(H)p_m \right]/\bar{f} \qquad (4.10)$$

This equation is known as the **schema theorem**, or the **fundamental theorem of genetic algorithms**. This is rather a grand title, but the success of the GA relies upon the proliferation of good schemata, and this equation tells us how rapidly this will occur; it is therefore a central equation in genetic algorithms.

4.7 Applications of the GA

Problems for which the power of the GA is useful are quite well defined; we can expect them to show three particular features:

● *The problem presents a significant challenge to conventional methods of solution.*

We have seen earlier that AI methods are not intended for use in simple problems, and this remains true for the GA, which becomes progressively more valuable as the scale of the problem grows, because of implicit parallelism. GA problems must therefore be difficult to solve; indeed, for some very large scale problems it appears that the GA is the only realistic method of attack presently available.

● *It must be possible to cast the solutions to a problem in the form of a string which the GA can manipulate.*

For some problems this cannot be done, and then the GA cannot be used. However, the definition of a 'string' is not quite as restricted as our discussion so far has suggested: recent work on the application of GAs to the dispersal of pollution suggests that for some problems, though a string may be inappropriate, a matrix description of the solutions may be used. This matrix is then manipulated in the usual way by the crossover, reproduction, and mutation operators.

• *Since the GA works by manipulating schemata, the GA will be unable to tackle a problem successfully unless a solution can be assembled by bolting together segments of strings.*

Suppose the 'correct' solution to a problem is coded by the string $\{\alpha_1, \alpha_2, \alpha_3, \alpha_4, \delta_5, \delta_6, \delta_7\}$. Strings containing schemata such as $\{\alpha_1, \alpha_2, \alpha_3, *, *, *, *\}$, or $\{*, *, *, *, \delta_5, *, \delta_7\}$, or $\{*, *, *, \alpha_4, \delta_5, *, *\}$ must generally have higher fitness than randomly-chosen strings, or the GA will not know how to select strings that contain good schemata.

These requirements – that the problem be complex, that solutions be codable in string form, and that it be possible to build solutions by bolting together good sub-strings – inevitably impose some restrictions on the range of problems to which the GA can be applied. Despite this, there are numerous problems in which the GA can be useful, and if a problem can be cast in this form, it may produce solutions of a quality and at a speed that is unattainable through any other current method.

We have already met the use of the GA in conformational analysis through our sample problem. This illustrative problem has been scaled up by many workers and the GA is now used to tackle the much more demanding protein-folding problem with some success. We shall illustrate the application of the GA in science with a further pair of examples.

Optimization of synthetic routes in organic chemistry

Chemists need to synthesize chemicals cheaply, efficiently and from readily-available precursors. The synthetic route to a complicated organic molecule is at present almost always devised by a human, but computer programs, whose purpose is to propose suitable synthetic routes to a desired product, are becoming more powerful and widely-used.

The choice of a suitable route for a synthesis of three or four steps is often a fairly routine matter, but if ten or twelve synthetic steps are required, much depends upon the experience of the chemist in the development of a high-yield route. There are probably millions of possible routes that use known synthetic steps, since at each stage there may be scores of reactions that each intermediate might undergo (Fig. 4.15).

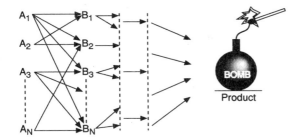

Fig. 4.15 An illustration of synthetic routes in organic chemistry.

Figure 4.15 is reminiscent of the chess-playing problem discussed in Chapter 1, and this is an example of a scientific problem which shows combinatorial explosion. Each GA string is chosen to represent a possible synthetic route to transform precursor A_n into the desired product. In this instance, while simple one-point crossover is a suitable mating operator, not every string can be validly crossed with every other string, since two strings can be crossed only when both have produced the same intermediate (Fig. 4.16).

$$\{A_1 \quad B_7 \quad C_4 \quad D_{11} \quad E_4 \quad F_{21} \quad G_6 \cdots\cdots Z\}$$

Suitable
crossing point

$$\{A_1 \quad B_9 \quad C_6 \quad D_{11} \quad E_7 \quad F_9 \quad G_9 \cdots\cdots Z\}$$

Fig. 4.16 A valid string crossover for the synthetic routes problem.

Applications of this sort show great potential in combination with organic synthesis expert systems.

Flowshop scheduling

In the industrial synthesis of chemicals, manufacturers look to benefit from economies of scale. However if a chemical plant is devoted to the production of a single high-value product such as a pharmaceutical or paint dye, high-volume production can be self-defeating if the chemical can be sold in only small quantities.

When synthesizing such chemicals, it is common to operate a 'chemical flowshop' (Fig. 4.17), in which different chemicals are synthesized within a single installation by passing the appropriate precursors into an array of reactors, dryers etc. whose configuration remains unchanged while different materials are produced.

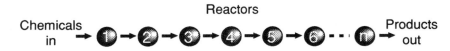

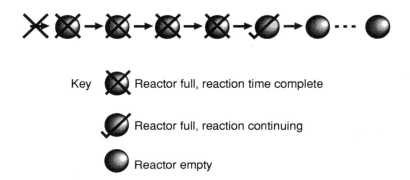

Fig. 4.17 A chemical flowshop. Different products can be synthesized in a single set of reactors by feeding in the appropriate precursors.

A prime difficulty with such an arrangement is that, if the plant manager does not choose the order in which products are made with care, the line can become temporarily blocked if reaction in one particular vessel continues while reaction in the preceding vessels is over (Fig. 4.18); this has an obvious negative effect upon the efficiency of the flowshop.

Fig. 4.18 A flowshop line containing a temporary block.

To minimize such blockages, it is important that chemicals be produced in an order such that intermediates can move in nearly 'lock-step' fashion through the flowshop (Fig. 4.19).

Fig. 4.19 Chemicals passing through a flowshop in 'lock-step' fashion.

If n chemicals are produced by the flowshop, the number of different orders in which those chemicals can be made is $n!$, so for a flowshop synthesizing 20 products, there are roughly 10^{18} different orders; in only a small proportion of these will the chemicals move from vessel to vessel in a nearly lock-step fashion. Since it is not possible to evaluate how good each of these 10^{18} orders is, some intelligent method must be used to find near-optimum orders.

A GA string for this problem consists of the order in which chemicals enter the first reactor. For example, the string {4, 18, 2, 14, 17, ...} means that chemical 4 is the first to enter the flowshop; when processing of this chemical in the first reactor is finished, it moves to the second reactor, allowing chemical 18 to enter the line, and so on. We note both that the size of the problem is large (10^{18} solutions), and that segments of strings can readily be interpreted in terms of schemata (corresponding to a sub-group of chemicals that move together through the flowshop in a partial lock-step fashion). Strings of this sort can readily be optimized using the GA and results from GA flowshop calculations show this to be a particularly powerful approach.

4.8 Improvements to the simple GA

The sections above outline a basic GA, which is effective for a wide range of problems. However, the performance of the algorithm can be improved by various adjustments, whose object is to speed convergence or reduce the chance of convergence to a sub-optimum solution. In this section we consider two widely-used enhancements to the algorithm.

Stochastic remainder

Although the roulette wheel mechanism biases the selection of strings towards the fittest, it may still fail to select the best string in a population for reproduction, which is clearly undesirable.

Stochastic remainder selection

1. Scale all strings so that the average fitness is 1.0
2. For every string with above average fitness, make a number of copies equal to the integer part of their fitness. Thus a string with a scaled fitness of 3.7 would have three copies placed into the next generation. The number of copies made is subtracted from the scaled fitness, so that all strings have a remainder < 1.
3. Use a problablistic procedure to select strings to fill the remaining spaces in the population. Choose a string at random and generate a random number. If the random number is less than the remainder of the fitness, then (a) make a copy of the string, and (b) reduce the fitness remainder to zero to prevent further copies of the string being made; if the random number is greater than the remainder, choose another string.
4. Continue until the total number of copies equals the population size.

This procedure ensures copies of the fittest strings will always be made into the next generation and can have a highly beneficial effect on the rate of convergence (Fig. 4.20).

Stochastic remainder selection is a strategy designed to ensure that all above-average strings will place at least one child in the new population, without unduly damaging the probablistic nature of the selection procedure.

The steps in stochastic remainder selection are shown above.

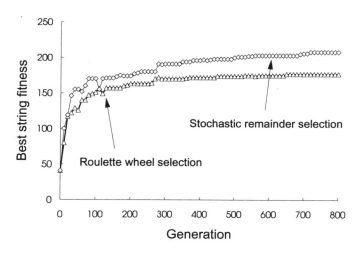

Fig. 4.20 The beneficial effects of stochastic remainder selection.

Elitism

Stochastic remainder guarantees that every string with above-average fitness will be copied at least once into the new generation, but once there, it is still possible for those strings to be destroyed by crossover. An elite procedure ensures that the best string in the current population is never lost.

One way of doing this, as a new population is being constructed, is to flag a copy of the fittest string and protect it from attack by crossover or mutation. If this strategy is combined with stochastic remainder to guarantee that the fittest string is always reproduced, the best fitness in the population can never decrease.

A second elitist method permits any string in the new population, including the best one, to suffer crossover. When crossover is complete, the fitness of every string in the new population is determined. If none of the strings in the new population is as fit as the best in the old, one of the least fit strings in the new population is displaced by the best of the old. This is a rather more powerful strategy than the first; crossover of the fittest string is potentially more valuable than crossover involving any other string in the new population, so protecting this from crossover, as the first elitist method does, is sometimes counter-productive.

Other techniques to enhance the progress of the GA may also be used, such as the retention of copies of good strings from earlier generations in a pool of 'useful genes'; these are then occasionally re-inserted into the population. Good schemata may be discovered, but sometimes accidentally lost as the algorithm proceeds, even using an elitist strategy. By re-introducing a few of the best strings from earlier generations perhaps once every hundred generations, we need not wait for the algorithm to rediscover good genetic material, and there is less likelihood that the calculation will become trapped at sub-optimal solutions.

These last few paragraphs seem to suggest that we can introduce additional or amended operators into the GA in what may appear to be an arbitrary fashion. Curiously, this is exactly the case: there is no fixed prescription according to which a GA must be run, though the basic operators of reproduction, crossover, and mutation always remain. Variables such as population size and crossover rate must be determined by the user, and there is further freedom to choose strategies such as stochastic remainder, or elitist selection. Both experience and experiment play a large part in the choice of productive strategies and suitable parameter values.

Postscript

Our ability to analyse and interpret scientific data is being significantly extended through the use of AI methods; this book has been able to provide just a hint of their power (but enough, we hope, to encourage you to investigate some of them further). As experience in their use expands, new roles and uses for them are being found. Their use is still in its infancy, but AI techniques are destined to be of central importance in science within a few years. The time when scientists use computers not just as tools to help in the analysis of data, but as partners in the development of scientific laws and understanding has almost arrived.

Further reading

Davis, L. (ed.) (1991). *Handbook of genetic algorithms*. Van Nostrand Reinhold, New York.
Goldberg, D.E. (1989). *Genetic algorithms in search, optimisation and machine learning*. Addison-Wesley, Reading, Mass.
Holland, J. (1975). *Adaptation in natural and artificial systems*. University of Michigan Press, Ann Arbor, MI.

Index